The Economics of Energy Efficiency

The Economics of Energy Efficiency

Barriers to Cost-Effective Investment

Steve Sorrell
SPRU: Science and Technology Policy Research, University of Sussex, United Kingdom

Eoin O'Malley
Economic and Social Research Institute, Ireland

Joachim Schleich
Fraunhofer Institute for Systems and Innovation Research, Germany

Sue Scott
Economic and Social Research Institute, Ireland

Edward Elgar
Cheltenham, UK • Northampton, MA, USA

Published by
Edward Elgar Publishing Limited
Glensanda House
Montpellier Parade
Cheltenham
Glos GL50 1UA
UK

Edward Elgar Publishing, Inc.
136 West Street
Suite 202
Northampton
Massachusetts 01060
USA

A catalogue record for this book
is available from the British Library

ISBN 1 84064 889 9

Printed and bound in Great Britain by MPG Books Ltd, Bodmin, Cornwall

Contents

Preface

In December 2003, representatives from most of the countries in the world gathered in Milan for the 9th Conference of Parties to the UN Framework Convention on Climate Change (UNFCCC). While the formal conference agenda was dominated by technical details, the political debate centred on the ratification of the Kyoto Protocol by the Russian Federation. Without Russian ratification, the Protocol will not enter into force and global efforts to mitigate climate change will be set back by as much as a decade.

The continuing uncertainty over Russian ratification has created difficulties for supporters of the Protocol, such as the European Union. On 15 December 2003, the EU energy commissioner Ms Loyola de Palacio claimed that it would be 'suicide' for the EU to continue to respect the Protocol without Russian participation. This brought a fierce rebuttal from the EU environment commissioner Margot Wallström who insisted that 'we can meet our Kyoto obligations without harming our economy'.

The costs and benefits of reducing carbon emissions are therefore a continuing focus of conflict at the highest political level. At the heart of this conflict lies a question which has been fought over within energy economics for nearly a quarter of a century, namely: are there widespread and cost-effective opportunities available to improve energy efficiency? The polar positions of the two EU commissioners reflect different answers to this question and each could point to analytical studies in support of their views. Hence, while the question may be 'old hat' to many energy economists, their discipline has failed to provide an adequate and universally acceptable answer. The 'barriers debate' is alive and well and continuing to provide a focal point of political controversy. Given the critical importance of this debate for the politics of climate change, there is clearly room for further empirical research. This book seeks to contribute to this ongoing debate and to build a bridge between the energy economics community and engineering-based practitioners in the energy efficiency field.

The book reports the results of a European Commission funded project *Barriers to Energy Efficiency in Public and Private Organisations (BARRIERS)*, which was conducted over the period 1998-2000 by researchers from SPRU (Science and Technology Policy Research) in the UK, the Fraunhofer Institute for Systems and Innovation Research (Fraunhofer ISI) in

Germany, and the Economic and Social Research Institute (ESRI) in Ireland. The authors gratefully acknowledge the financial support of the Commission, as well as the guidance from Birgit Bauman and Pekka Jarvilehto of DG Research.

The book has its origins in a research proposal developed at SPRU during the summer of 1997 by Professor Jim Skea (now Director of the Policy Studies Institute, London), Rupert Howes (now Director of the Sustainable Economy Programme, Forum for the Future, London) and Steve Sorrell (project co-ordinator). Particular thanks are due to Jim Skea who conceived the project idea and directed the project during its early stages. Thanks also to Professor John Chesshire of SPRU for taking over from Jim as the Authorised Scientific Officer in 1999 and for sharing his wealth of knowledge of energy efficiency.

The constraints of contract research have led to an uncomfortably long delay between completing the BARRIERS project and finalising the manuscript for this book. The project reports were published in November 2000 and the authors have spent much of the intervening period immersed in the fast moving world of climate policy and emissions trading. Thanks are due to SPRU for providing three months salary support to Steve to complete the book and to Dymphna Evans and colleagues at Edward Elgar for their patience.

Important contributions to the empirical research for the project were made by Ulla Boede, Dirk Koewener, Wilhelm Mannsbart, Katrin Ostertag and Peter Radgen of Fraunhofer ISI, and by Fergal Trace and Fergal Shortall of the ESRI. Katrin Ostertag provided some valuable insights into the role of transaction costs and energy efficiency, and her book on this topic – *No-Regret Potentials in Energy Conservation: An Analysis of their Relevance, Size and Determinants* – has recently been published by Physica-Verlag, Heidelberg.

The project benefited considerably from the input of an Advisory Group, consisting of: Peter Horrocks (European Commission); Ralph Lennartz (European Commission); John Huddleston (Future Energy Systems); John Newman (International Energy Agency); Richard Paton (UK Department of Trade & Industry); David Somervell (University of Edinburgh); Mike Spillane (Brewers and Licensed Retailers Association); and David Taylor (Irish Energy Centre). Particular thanks are due to David Somervell for his valuable advice and insights into energy management practices in universities, and to Mike Spillane of the former Brewers and Licensed Retailers Association (now the British Beer and Pub Association) for his assistance with the energy survey of UK breweries.

This project would not have been possible without the co-operation of the many individuals in the case study organisations who gave up their valuable time to complete questionnaires and be interviewed. Each contributed greatly to our understanding of the topic with informative and vivid accounts of their experiences. Since a key finding from the project was the importance of staff time constraints in each sector, their co-operation is particularly appreciated. Thanks also to the interviewees from energy service companies in each country and to the respondents to the postal surveys of energy management practices in UK universities and UK breweries.

The authors have benefited from insights and comments from a number of sector specialists including Peter Viebahn from eco-campus net; Georg Benke from the Austrian Energy Agency (EVA); Professor David Gann and Professor James Barlow of Imperial College; Andrew Parish and colleagues at the former Irish Energy Centre (now Sustainable Energy Ireland); and attendees at a UK Energy Share Fair for university energy managers. They have also benefited from the comments of colleagues, including Frans Berkhout, Erik Millstone, Nick von Tunzelmann, Adrian Smith, Julia Hertin, and Christina Volkers (all of SPRU). Finally, thanks to Angeliki Panteli for her love, support and encouragement during this lengthy task.

Steve Sorrell (Project Coordinator)
Brighton, December 2003

Abbreviations

AEAT	AEA Technology
BEMS	Building Energy Management System
BLRA	Brewers and Licensed Retailers Association
BREEAM	Building Research Establishment Environmental Assessment Methodology
BWTE	Brewers Without Tied Estates
CCA	Climate Change Agreement
CCL	Climate Change Levy
CEMEP	European Committee of Manufacturers of Electrical Machinery and Power Electronics
CGE	Computable General Equilibrium
CHP	Combined Heat and Power
CVHR	Copper Vapour Heat Recovery
DCF	Discounted Cash Flow
DETR	Department of the Environment, Transport and the Regions
EEBPP	Energy Efficiency Best Practice Programme
EMS	Environmental Management Scheme
ENUSIM	Energy Use Simulation Model
ESCO	Energy Services Company
EU ETS	EU Emissions Trading Scheme
FES	Future Energy Systems
GDP	Gross Domestic Product
GHG	Greenhouse Gas
GJ	GigaJoule
GW	GigaWatt
GWh	GigaWatt hour
HE	Higher Education
HEFCE	Higher Education Funding Council for England
hl	hectolitre
HOAI	Ordinance of Fees for Architects and Engineers (German)
HVAC	Heating, Ventilation and Air Conditioning
IEC	Irish Energy Centre
IPCC	Intergovernmental Panel on Climate Change
IPPC	Integrated Pollution Prevention and Control

IPSEP	International Programme for Sustainable Energy Paths
IRR	Internal Rate of Return
IT	Information Technology
JPPSG	Joint Procurement Policy and Strategy Group
M&T	Monitoring & Targeting
MACC	Making a Corporate Commitment
MD	Managing Director
MJ	MegaJoule
NAO	National Audit Office
NGO	Non-Governmental Organisation
OECD	Organisation of Economic Co-operation and Development
PFI	Private Finance Initiative
PIQ	Pre-Interview Questionnaire
PJ	PicaJoule
SEC	Specific Energy Consumption
SME	Small and Medium-sized Enterprises
TCE	Transaction Cost Economics
TJ	TeraJoule
UK ETS	UK Emissions Trading Scheme
VM	Value Management
VSD	Variable Speed Drive
WACC	Weighted Average Cost of Capital
WLC	Whole-Life Costing

1. Introduction

Steve Sorrell

1. THE BARRIERS DEBATE AND THE POLITICS OF CLIMATE CHANGE

In 1997, the US House of Representatives asked the Energy Information Administration (EIA) to examine the potential impacts of the Kyoto Protocol on the US economy. The results of this study, which was published in October 1998, were dramatic (EIA, 1998). The EIA examined a scenario in which the US would meet its Kyoto obligations solely through the abatement of carbon dioxide (CO_2) emissions, with no reduction of other greenhouse gases (GHGs) and no use of international emissions trading or carbon sinks. In this scenario, US GDP was reduced by 4.2 per cent from the reference case, average electricity prices increased by 86 per cent, gasoline prices increased by 53 per cent and coal consumption reduced by 77 per cent. The chairman of the Committee of Science of the US House of Representatives commented that: 'We now know the real story behind the Kyoto Protocol: energy use will be more expensive, economic growth will be jeopardised and American families will pay dearly for a flawed treaty' (Sensenbrenner, 1998). Two years later, President Bush withdrew the US from the Kyoto Protocol, citing flaws in the science, the exclusion of 80 per cent of the world's population and the serious negative impacts on the US economy. While this decision has been derided throughout the world, the EIA report was one of a number of studies that gave analytical backing to the President's stance.

In July 2001, another major study on the impact of the Kyoto Protocol on the US economy was published by the International Project for Sustainable Energy Paths (IPSEP) (IPSEP, 2001). This combined the results of a number of studies, including the influential *Scenarios for a Clean Energy Future* (CEF), prepared by researchers from five national laboratories of the US Department of Energy (Interlaboratory Working Group, 2000; Brown et al., 2001). In stark contrast to the EIA findings, the IPSEP study suggested that the US could meet its Kyoto obligations by 2010 with a net *gain* in economic output of $60 billion/year, or 0.5 per cent of GDP (increasing to 1 per cent by 2020). Furthermore, this would be accompanied by an 8 per cent reduction in the total national energy bill.[1]

The sharp differences between the EIA and IPSEP studies are emblematic of a wider debate within energy economics. Despite a history stretching back more than 25 years, this debate remains largely unresolved. In simple terms, the debate is between one group of researchers using 'top-down' or macroeconomic modelling tools who find there are significant costs associated with reducing energy use and carbon dioxide emissions, and a second group using 'bottom-up' or engineering-economic modelling tools who argue that there are substantial opportunities to mitigate climate change that will allow emission targets to be achieved at negative cost (Box 1.1). This debate involves disagreements over the methodologies and concepts appropriate to studying economic behaviour, the validity of various modelling assumptions and the appropriateness of different modelling tools (Canes, 2002; Diesendorf, 1998; Grubb et al., 1993; IPCC, 2001; Laitner et al., 2003; Wilson and Swisher, 1993), But while the debate has a technical character, the results can be enormously influential in shaping political debates on climate change. This is particularly apparent in the case of the US, whose actions have fundamentally weakened the Kyoto Protocol and seriously delayed international action to reduce greenhouse gas emissions (Den Elzen and De Moor, 2003).

Box 1.1 Top-down versus bottom-up approaches to energy modelling

Cost-effective abatement opportunities are commonly identified by 'bottom-up' or engineering-economic models (Interlaboratory Working Group, 1997; Marsh et al., 2003).[2] These usually focus upon individual sectors and energy services and contain detailed information on the performance and cost of energy-using equipment. They simulate the ageing and replacement of this equipment and seek to minimise the net cost of energy services given appropriate assumptions about individual and organisational investment behaviour. These models typically assume that transaction costs are negligible and ignore the feedbacks between energy decisions and the rest of the economy. The attractive claim of such models is that both carbon emissions and money can be saved at the same time, with the results frequently suggesting that profitable investments with paybacks as short as two years are widely available. Several empirical studies of real-world energy decision-making appear to support this claim (Romm, 1999; Lovins and Lovins, 1997).

In opposition to this view, many economists claim that the markets for energy and energy-using technology are broadly efficient. If consumers or organisations are employing high discount rates they may be responding rationally to high risk investments, or there may be a range of 'hidden costs' of technology adoption which are not captured by bottom-up models. This perspective is supported by the results of some macroeconomic, or top-down, models which operate at a high level of aggregation and represent technology through abstract concepts such as production functions (EIA, 1998; Weyant and Hill, 1999; Nordhaus, 2000).[3] Top-down models include the feedbacks between the energy sector and other sectors of the economy and typically assume that the economy is operating close to the competitive equilibrium. Price and income elasticities estimated from historical data sets are assumed to incorporate complex behavioural factors such as the response to hidden costs, with the implication that any real opportunities to save money will have already been taken. A consequence of this assumption is that reductions in greenhouse gases can only be achieved at the expense of reductions in the output of other goods and services. Critics of the top-down approach argue that this assumption simply rules out the possibility of 'no regrets' opportunities (DeCanio, 2003; Sanstad et al., 2001; Wilson and Swisher, 1993).

The gap between these two approaches is beginning to narrow. Hybrid models have been developed which incorporate elements of both bottom-up and top-down approaches, and considerable progress has been made in incorporating technological change (Grubb and Kohler, 2000; Löschel, 2002). But despite these developments, conflicting modelling results still dominate climate policy debates. For example, the UK government's reliance upon a bottom-up model (MARKAL) to assess the cost implications of a 60 per cent reduction in GHG emissions has drawn strong criticism from orthodox economists (Helm, 2003), while the industry-backed International Council for Capital Formation (ICCF) has used the results of top-down studies to criticise official estimates of the cost of meeting the Kyoto targets (ICCF, 2002).

The potential for cost-effective, or 'no-regrets'[4] opportunities to reduce GHG emissions has been recognised by the Intergovernmental Panel on Climate Change (IPCC):[5]

> '…Half of the potential emission reductions may be achieved by 2020 with direct benefits (energy saving) exceeding direct costs (net capital, operating and maintenance costs).' (IPCC, 2001, p. 40)

The great majority of these opportunities are for energy efficiency improvements across all end-use sectors. These frequently involve the adoption of established technologies whose performance is well proven and which involve little technical risk. But if these no regrets opportunities are already available, why are they not being taken up? What is the nature of the 'barriers' which prevent individuals and organisations from investing in technologies which both save money and reduce emissions? Can these barriers be overcome by suitable forms of policy intervention? If so, will the costs of such intervention outweigh the benefits to be achieved?

These are familiar questions and the subject of a considerable literature (Brown, 2001; Golove and Eto, 1996; IPCC, 2001; Jaffe et al., 1999; Krause, 1996). But despite the number and range of studies, a consensus has yet to emerge. While modelling studies simply assume answers to these questions, empirical studies are inconsistent in their results and comprehensive policy evaluations are difficult to find. The result is a diversity of views on the costs of greenhouse gas mitigation, the nature and importance of different barriers and the likely success of policy interventions (Golove and Eto, 1996; Sutherland, 1996; Joskow and Marron, 1992; Eto et al., 2000).

The further development of modelling techniques can only go so far in resolving these disputes. Ultimately what is required are detailed empirical studies of the nature, origin and operation of the supposed barriers to energy efficiency in a wide range of energy service markets, together with evaluation studies of the costs and benefits of different types of policy intervention. The IPCC (2001, p. 71) recommend *measuring* these barriers, but a more realistic objective would be to assess the relative importance of different barriers in different contexts and their aggregate impact on energy and economic efficiency. To verify the claims of bottom-up modellers it would be necessary to demonstrate that first, the barriers explain the non-adoption of cost effective energy efficient technologies; and second, that a substantial proportion of these barriers could be cost-effectively overcome through non-price regulatory policies.

This book addresses these issues through a comprehensive examination of energy management practices within a wide range of public and private sector organisations. The aim is to determine whether and to what extent such

organisations are neglecting cost-effective opportunities to improve energy efficiency, the reasons for this, and the options that are available to encourage these opportunities to be taken up. In doing so, the book seeks to test the usefulness of different economic theories in explaining organisational behaviour in relation to energy efficiency and to assess the implications of these findings for public policy. It is hoped that the research will shed new light on the barriers debate and provide a useful input to the analysis and development of climate policy.

2. RESEARCH DESIGN

2.1 Objectives and Research Questions

This book is based upon a two-year research project, *Barriers to Energy Efficiency in Public and Private Organisations* (BARRIERS), which was conducted by research teams in the UK, Ireland and Germany over the period 1998–2000.[6] The aim of the project was to examine how barriers to energy efficiency operate, the circumstances in which they arise and the extent to which public policy or organisational change may overcome them. The research teams conducted a total of 48 case studies of energy-using organisations in the brewing, mechanical engineering and higher education sectors, supplemented by postal surveys, documentary analysis and contextual interviews with relevant actors. In each case, the unit of analysis was the decision-making procedures within the individual organisation focusing upon those decisions which had an important influence on overall energy efficiency. In particular, the project sought to identify:

- the extent to which the selected organisations were neglecting cost-effective opportunities to improve energy efficiency;
- the reasons for this, including the existence, nature, operation and determinants of any barriers to energy efficiency;
- the relative and aggregate importance of these barriers for different types of decision within different organisational contexts; and
- the potential for organisational, contractual and public policy measures to overcome these barriers, including the extent to which the identified barriers legitimate policy intervention.

The research assessed the extent to which the case study organisations had adopted a number of existing and well-proven energy efficient technologies that are promoted in best practice literature and have the potential for providing energy services at a lower net cost. But while investment decisions for these technologies are the principal focus of the barriers debate, these are not the only determinants of organisational performance in energy efficiency. In practice, organisational performance is influenced by a wide range of internal and external factors which also need to be considered.

First, the organisation lies at one end of a complex supply chain for energy services, and the decisions made by a network of intermediaries can have a determinate impact on energy performance and leave little room for choice by the final user. For example, the energy consumption of a lathe will largely be determined by the choice of motor by the lathe supplier, while the energy consumption of a building will be largely determined by its architectural design and the decisions made by equipment suppliers and building contractors. An alternative research design would be to examine the contribution of different actors along these supply chains for particular categories of energy service, such as motive power (de Almeida, 1998). But the approach taken in this project was to examine decision-making by the end-user organisation and to identify the barriers to energy efficiency from an end-user perspective.

Second, investment in technologies which are specifically designed to improve energy efficiency (such as thermal insulation) is likely to be the exception rather than the rule. For most types of investment, energy efficiency will be just one aspect of a technology's performance and is unlikely to be either the primary motivation for the investment or the primary purpose of the technology. For example, a new motor may be purchased to replace an existing motor which has broken down. While both energy efficient and inefficient choices may be available, this is likely to be a secondary consideration to the purchaser, who is primarily concerned about price, technical compatibility and availability. In general, the energy consumption of an organisation will be largely determined by the performance of its inherited building and capital stock, and the energy efficiency opportunities when major elements of this stock are expanded or replaced may significantly exceed those from retrofitting specific energy efficient technologies. Hence, decision-making on new plant and buildings is of fundamental importance in determining overall organisational performance.

Third, organisational performance in energy efficiency will be strongly influenced by the procedures for operating and maintaining equipment. For example, simple 'housekeeping' procedures, such as checking for compressed air leaks, are widely acknowledged to be highly cost effective, and variations in equipment operation and maintenance patterns can lead to wide variation in energy consumption, even where similar technologies are involved (Energy Efficiency Office, 1993).

Hence, organisational performance in energy efficiency will be determined by a wide range of factors and influenced by a wide range of organisational decisions. In view of this, the project sought to describe, understand and explain the particular factors which inhibited organisations from taking decisions which could cost-effectively improve aggregate energy efficiency. This led to the following research questions:

- Why do organisations impose very stringent investment criteria for projects to improve energy efficiency?
- Why do organisations neglect projects that appear to meet these criteria?
- Why do organisations neglect energy efficient and apparently highly cost-effective alternatives when making broader investment decisions?
- Why do organisations neglect energy efficient and apparently highly cost-effective alternatives when making operational, maintenance and purchasing decisions?
- How can the neglect of cost-effective opportunities be overcome in these different types of decision, and what is the appropriate role of the host organisation, other private sector actors and public policy?

These questions gave the project a broad scope. Instead of asking why a single type of energy efficient technology (for example, high frequency fluorescent lighting) had or had not been adopted, the project sought to determine why a broad range of decisions had or had not been taken. Similarly, instead of focusing solely upon the adoption of specific energy efficient technologies, the project explored the full range of reasons why organisations performed better or worse in energy efficiency. The dependent variable was therefore 'organisational performance in energy efficiency' and this was assessed in a variety of ways, including:

- quantitative measures of organisational energy consumption (where available), normalised to suitable measures such as floor area or production output and compared with sector-specific benchmarks;
- responses to self-assessment questionnaires on organisational performance in energy management;
- documentary evidence, including energy/environmental reports, the results of energy audits and recorded trends in energy consumption;
- responses to interview questions on energy management practices, investment opportunities, barriers to energy efficiency and other factors; and
- quantitative data on the extent of adoption of a number of energy efficient technologies, selected from best practice literature and regarded as available, reliable and cost-effective.

Hence, while decisions to adopt (or not adopt) specific energy efficient technologies were an important focus of the empirical research, they were by no means the only item of interest. The objective within each case study was to judge the overall performance of the organisation in relation to energy efficiency and to determine the reasons for this. This was achieved by combining multiple sources of quantitative and qualitative evidence and seeking corroboration and convergence in this evidence. The reason for choosing this approach in preference to econometric methods was that organisational performance in energy efficiency is difficult to measure and the factors that influence this performance are difficult to isolate through survey methods or to represent as simple variables. For example: quantitative measures of organisational energy consumption may be unavailable, and where they are available it may be difficult to normalise them to an appropriate output measure or to obtain suitably disaggregated sector-specific benchmarks. Even where benchmarks are available, factors such as the size and age of the plant or the particular mix of products may make simple comparisons with these benchmarks problematic.

The advantage of this multidimensional approach is that it gives a much richer picture of organisational decision-making on energy efficiency, including the different factors which influence this decision-making, the diversity of barriers which are faced and the range of policies which may be required to overcome them. But the disadvantage is that it makes the testing of individual hypotheses more difficult, since the relative importance of each barrier may be expected to vary between different types of organisational decision.

2.2 Hypotheses and Operational Measures

To develop the research questions further, it was necessary to state some hypotheses that identify the factors that are likely to be relevant. These are the proposed barriers to energy efficiency, which are defined as postulated mechanisms that inhibit a decision or behaviour that appears to be both energy efficient and economically efficient. A key component of the project was the development and elaboration of a taxonomy of barriers, drawing upon ideas from orthodox, transaction cost and behavioural economics. This taxonomy was based upon categories that are widely used within the energy efficiency literature, but the description and evaluation of these barriers drew upon more formal ideas, such as asymmetric information. A summary of the proposed taxonomy is provided in Table 1.1, while a comprehensive discussion is provided in Chapter 2. Each of the proposed barriers may be considered a hypothesis that potentially explains the neglect of energy efficiency within organisational decisions. But as discussed in Chapter 2, each of these barriers may have a number of contributory mechanisms, and several of these mechanisms may coexist in different situations. For example, imperfect information on energy efficiency opportunities may result from both the nature of information as a public good and the overhead costs of establishing and maintaining an organisational energy information system. Hence, the aim of the project was both to identify the relative importance of the six barriers listed in Table 1.1 and to identify the contribution of different mechanisms towards these barriers.

The next step was to identify a number of measures that would allow the existence and importance of the hypothesised barriers to be determined from the case study research. As with the dependent variable, this involved assembling information from a wide range of sources and seeking replication and corroboration. For example, the importance of split incentives was explored by mapping the responsibilities for capital and operational expenditures within each organisation, identifying the perceived strength of incentives in each case, and assessing whether there were differences in performance between those cases where individuals and departments were accountable for energy costs and those where they were not. In each case, the primary source of evidence was semi-structured interviews with a wide range of individuals from the case study organisations, including energy/environment managers, financial managers, departmental managers, members of energy/environment committees, and engineering and maintenance staff. This was supplemented by documentary evidence and the use of pre-interview questionnaires which included detailed questions on energy management practices, the adoption of specific technologies and the perceptions of barriers to energy efficiency.

2.3 Selection of Case Studies

The project focused upon energy management in three classes of organisation:

- *Higher education institutions:* These are representative of public sector organisations where energy accounts for less than 2 per cent of total costs and where energy use is dominated by building service applications, such as heating, lighting and air conditioning.

Table 1.1 A taxonomy of barriers to energy efficiency

Barrier	**Claim**
Risk	The short paybacks required for energy efficiency investments may represent a rational response to risk. This could be because energy efficiency investments represent a higher technical or financial risk than other types of investment, or that business and market uncertainty encourages short time horizons.
Imperfect information	Lack of information on energy efficiency opportunities may lead to cost-effective opportunities being missed. In some cases, imperfect information may lead to inefficient products driving efficient products out of the market.
Hidden costs	Engineering-economic analyses may fail to account for either the reduction in utility associated with energy efficient technologies, or the additional costs associated with them. As a consequence, the studies may overestimate energy efficiency potential. Examples of hidden costs include overhead costs for management, disruptions to production, staff replacement and training, and the costs associated with gathering, analysing and applying information.
Access to capital	If an organisation has insufficient capital through internal funds, and has difficulty raising additional funds through borrowing or share issues, energy efficient investments may be prevented from going ahead. Investment could also be inhibited by internal capital budgeting procedures, investment appraisal rules and the short-term incentives of energy management staff.
Split incentives	Energy efficiency opportunities are likely to be foregone if actors cannot appropriate the benefits of the investment. For example, if individual departments within an organisation are not accountable for their energy use they will have no incentive to improve energy efficiency.
Bounded rationality	Owing to constraints on time, attention, and the ability to process information, individuals do not make decisions in the manner assumed in economic models. As a consequence, they may neglect opportunities for improving energy efficiency, even when given good information and appropriate incentives.

- *Brewing firms:* These are representative of process industries where energy accounts for less than 5 per cent of total costs and where energy use is dominated by process applications that are frequently specific to the individual sector.
- *Mechanical engineering firms:* These are representative of light industry where energy accounts for less than 2.5 per cent of total costs and where energy use is dominated by generic technologies, such as mechanical drives, furnaces and space heating.

In all three cases, energy forms a relatively small proportion of total costs and may easily be given a low priority within organisational decision-making. But these types of organisation account for around 20–25 per cent of final energy demand in OECD countries and studies regularly suggest that cost-effective opportunities to improve energy efficiency in these sectors are widely available (IPCC, 2001). Hence, if substantial reductions in greenhouse gas emissions are to be achieved, significant improvements must be achieved in the energy efficiency of these types of organisation.

The empirical work was conducted by teams in the UK, Ireland and Germany. The UK and Ireland form two ends of a spectrum in terms of the degree of energy market liberalisation, while Germany and Ireland contrast strongly in their reduction targets for greenhouse gas emissions.[7] The policy instruments used to encourage energy efficiency also vary widely between each country, with differing emphases being given to information campaigns, energy taxation, negotiated agreements, investment subsidies and support for energy audits. The research aimed to identify both similarities between energy-using organisations in these three countries and also differences resulting from the structure of the sector in each country, the relevant policy framework and other factors.

A total of 48 energy-using organisations were studied – four to seven organisations per sector per country. Table 1.2 summarises the distribution of case studies between each sector and country.

Table 1.2 Number of case studies in each country and sector

Sector	UK	Ireland	Germany	Total
Higher education	6	5	6	**17**
Brewing	5	5	5	**15**
Mechanical engineering	5	7	4	**16**
Total	**16**	**17**	**15**	**48**

It was difficult to achieve comparability between the organisations selected as case studies in each country because of differences in the national contexts. For example, Ireland had few large mechanical engineering firms and no small breweries. Instead, the individual cases were chosen to give a representative selection of the range of organisations found in each country and sector, using the criteria summarised in Table 1.3.

Table 1.3 Criteria for selection of case studies

Higher education	• at least one campus university • at least one metropolitan university • at least one small university (utilities bill <£1m/year) • at least one large university (utilities bill >£2m/year)
Breweries	• at least one large brewery (>1500 th hl/year) • at least one small brewery (<100 th hl/year) • at least one brewery owned by a major company • at least one independent brewery
Mechanical engineering	• at least one large firm (>500 employees) • at least one small firm (<250 employees) • at least one firm that is a subsidiary of a larger multinational

2.4 Conduct of the Research

The first stage of the research was to characterise the existing policy environment for energy supply and the provision of energy services within each country, together with the technical and economic situation of the case study sectors. This included:

- A review of energy markets in each country and the identification of relevant policy measures that promote energy efficiency. Climate policy was in a state of flux during the course of the project since each country was developing strategies to meet its Kyoto obligations.
- An analysis of the case study sectors within each country in terms of size, ownership, market share, technologies employed, recent economic performance, technological and market trends, policy influences and the prospects for growth or contraction.
- A review of the energy use patterns within the case study sectors, including sector benchmarks, historical trends, the split between different end-uses and the potential for cost-effective energy efficiency improvement. This included developing a list of cost-effective energy efficient technologies

which appeared appropriate for either retrofit or replacement applications within the sector.

The second stage was to develop concepts and protocols for conducting the empirical research. First, the taxonomy of barriers was developed with reference to ideas from orthodox, transaction cost and behavioural economics (Sorrell et al., 2000). Second, a range of quantitative or qualitative measures were suggested for each barrier which would allow its existence to be identified from case study research (Sorrell, 1999a). Third, standardised questionnaires and interview protocols were developed for a number of different interviewees within each sector (Sorrell, 1999b). These were based on the identified measures and were designed to ensure commonality of approach between the country teams. Examples of the questionnaires and interview protocols are provided in Annexes A and B.

The third stage was to identify and contact a number of organisations within each country and sector, and to secure their co-operation with the assistance of the relevant sector associations. Preliminary documentary information was obtained and the energy manager (or equivalent) was asked to complete a brief questionnaire containing questions on energy consumption, energy management practices and the adoption of specific technologies. A number of detailed on-site interviews were then conducted with the energy manager and other members of staff and additional documentary evidence obtained. The number of interviewees ranged from two within small mechanical engineering firms up to eight within large universities. Follow-up telephone interviews were then used to resolve ambiguities and obtain additional information, and a number of additional interviews were conducted with regulators, consultants and sector specialists.

The results from each country and sector were then brought together in nine country/sector reports, which included a characterisation of the sector, descriptions of the energy management practices within the case study organisations, a discussion of the evidence for different types of barrier, and an assessment of policy implications. Individual barriers were identified as being of either high, medium or low importance, and relevant measures which may overcome these barriers suggested.

In addition to the case studies, four additional pieces of research were undertaken:

- *UK brewery energy management survey*: A postal survey was undertaken of energy management practices within the UK brewing industry. The survey was conducted with the assistance of the UK Brewers and Licensed Retailers Association (BLRA) and was based very closely on the pre-interview questionnaire (Annex A). A total of 40 replies were obtained, representing 53 per cent of the survey population.

- *UK university energy management survey*: A similar survey was conducted of energy management practices within UK universities. A total of 32 replies were obtained, representing 33 per cent of the survey population.
- *Barriers to energy efficiency in the UK construction industry*: The energy efficiency of new buildings was a dominant theme of the university case studies and each case study provided a series of examples of cost-effective opportunities being lost. The reasons for this were varied, but derived primarily from the nature and operation of the construction industry and the associated mechanisms for project financing. In view of the importance of this issue, some additional research was undertaken to explore these issues in more detail, focusing primarily on building services.
- *The potential for energy service companies*: The outsourcing of energy management to an energy services company (ESCO) was identified as a promising route to overcoming organisational barriers to energy efficiency (James and Hopkinson, 2002; Hansen and Weisman, 1998). ESCOs offer a single contract to minimise the total bill for the services that energy provides, such as heating, lighting and air conditioning. This in turn allows the host company to lower risk, avoid capital expenditure, minimise transaction costs and concentrate attention on the core business. In view of the importance of this market, some additional research was undertaken to explore the potential for ESCOs within each of the countries and sectors.

2.5 Methodological Issues

Case studies are widely used within sociology and political science, but are treated with some scepticism within economics. Hence some justification for this approach is required.

In a classic textbook on the case study method, Yin (1994) defines case studies as being suitable when: '…a "how" or "why" question is being asked about a contemporary set of events over which the investigator has little or no control' (Yin, 1994, p. 9). Furthermore, a case study: '…investigates a contemporary phenomenon within its real-life context, when the boundaries between phenomenon and context are not clearly evident' (Yin, 1994, p. 13). The major comparison here is with a survey of a large number of organisations. The drawback with surveys for the present set of research questions is that their ability to investigate the context of organisational decision-making is limited. This means the survey designer must limit the number of variables to be analysed and the number of questions which may be asked. Furthermore, many of the most relevant variables are difficult to identify through simple questionnaire responses and difficult to incorporate within a quantitative model. This creates the risk that important factors will be inadequately represented or ignored. Case studies, in contrast can accommodate complex, multivariate situations in which the perspectives of a wide range of actors are relevant. Case studies allow a large number of variables to be explored and allow multiple sources of quantitative and qualitative evidence to be employed. In particular, case studies provide a means to identify causal mechanisms in real-life situations that are too complex for a survey and not amenable to controlled experimentation (DeCanio, 1994).

A common criticism of case studies is that they do not allow results to be 'generalised' to a wider population. But this criticism rests upon an inappropriate analogy with statistical methodologies. Multiple case studies should not be treated as statistically significant 'samples' of a population, but as attempts to either obtain similar results or obtain contrasting results for predictable reasons (Yin, 1994, p. 45). The appropriate analogy is with multiple experiments rather than multiple samples, since the objective is to *explain* the observed phenomenon through the identification of relevant causal mechanisms, rather than infer the distribution of characteristics within a population (Yin, 1994, p. 46).[8]

Yin (1994, pp. 34–5) suggests two criteria for assessing the quality of a case study design:

- *construct validity*: establishing correct operational measures for the concepts being studied; and
- *internal validity*: establishing a causal relationship, whereby certain conditions are shown to lead to other conditions, as distinguished from spurious relationships.

Yin's primary recommendation for establishing construct validity is the use of multiple sources of evidence which provide convergent lines of inquiry – that is, with the majority of evidence pointing in the same direction (Yin, 1994, pp. 90–4). But a logically prior step is the identification of the relevant concepts, such as barriers. This involves a process of abstraction, in which significant elements of the situation are identified, irrelevant elements are ignored and causal mechanisms are attributed on the basis of theoretical arguments regarding the nature of individual agents and the consequences of their relations and interactions – as embodied in specific organisational and institutional structures (Sayer, 1992; Lawson, 1997).[9] For example, the take-up of energy efficiency opportunities may be expected to depend upon the extent to which individuals/departments/organisations can appropriate the resulting cost savings. The importance of abstraction is often neglected in textbooks on methodology, but if the proposed concepts are not coherent, the research is unlikely to be productive. For this reason, the present study gives a great deal of attention to defining and understanding the different mechanisms which may contribute to the organisational neglect of energy efficiency (see Chapter 2). This includes the incentives and constraints provided by different organisational arrangements and the resources, preferences and decision-making capacity of the relevant actors.

Establishing internal validity requires the successful demonstration of causality. It is assumed here that causality cannot be established by the identification of simple regularities of the form: 'if X is present then Y will result'. This is because in a complex social system, the operation of any proposed causal mechanism may be inhibited by other countervailing mechanisms. For example, the tendency to neglect energy efficiency opportunities due to problems of split incentives may be inhibited by the interest and motivation of a particular CEO (Lawson, 1997).[10] This means that causality cannot be simply established or disproved by linking the apparent existence of a particular mechanism with the existence or otherwise of particular concrete outcomes. Instead, causality may be suggested by establishing as far as is possible (Runde, 1998):

- Whether the proposed causal mechanisms were *present*. For example, were the agents able to appropriate the benefits of energy efficiency investments?
- Whether the proposed mechanisms were *inhibited*. For example, did alternative incentive structures circumvent this problem?
- Whether the proposed mechanisms have sufficient *depth*. For example, why did the split incentive problem exist? What factors contributed to maintaining it? Can the split incentive problem be expressed in terms of more 'fundamental' mechanisms, such as excessive transaction costs?
- Whether the proposed mechanisms were *effective*. For example, would the investments still have been neglected even if the agents were able to appropriate the benefits?

The last step is particularly difficult, since the neglect of energy efficiency opportunities may in some cases be co-determined by the coexistence of multiple causal mechanisms. For example, the consequences of imperfect information on energy efficiency opportunities may be reinforced by split incentive problems. One possibility is that a single cause is dominant and the others have only a limited impact on observed behaviour, while another is that each cause is relevant and their net effect equals or exceeds the sum of their individual effects. Measures of the relative importance of different factors may be obtained through interviews or questionnaires, while the use of multiple case studies may allow the contribution of individual factors to be isolated. But the possibility remains that the neglect of energy efficiency opportunities is overdetermined. A comprehensive account would then require a synthesis of these different factors.

3. PLAN OF THE BOOK

The book is divided into three parts: an extended essay on understanding barriers to energy efficiency (Chapter 2); a selection of the empirical results from the project (Chapters 3 to 7); and a summary of the project conclusions (Chapter 8).

Chapter 2 sets out the theoretical framework for the project in some detail. The main argument of this chapter is that orthodox economic theory is necessary but insufficient for understanding barriers to energy efficiency. Understanding may be enhanced by introducing ideas from transaction cost economics and behavioural economics, and these ideas may also legitimate a wider scope for policy intervention.

The chapter begins by describing the history of the barriers debate and distinguishing between barriers and orthodox market failures. It then reviews the assumptions of orthodox theory and introduces the types of market failure relevant to the energy efficiency gap, focusing in particular on asymmetric information. It does the same for transaction cost and behavioural economics, identifying the nature and determinants of transaction costs, the importance of bounded rationality and the insights offered by social psychology. The chapter then draws upon ideas from each of these approaches to explore the nature and determinants of the six barriers to energy efficiency listed in Table 1.1. These are illustrated with concrete examples from energy service markets and energy-using organisations. The chapter then summarises the proposed typology and discusses whether and how these barriers may provide legitimate grounds for government intervention.

Chapters 3 to 6 provide a selection of the empirical results from the project, focusing upon those country/sector studies for which the results proved particularly illuminating. These were the German higher education sector, the UK higher education sector, the UK brewing industry and the Irish mechanical engineering industry. In each case, the reported results had much in common with the sector studies in the other countries, but also showed some interesting differences. These similarities and differences are discussed further in Chapter 8.

The discussion in Chapters 3 to 6 is split into four sections. First, the key features of the sector within the particular country are described, including ownership, size, structure, economic trends, patterns of energy use and opportunities for energy efficiency improvement. Second, the results from the individual case studies are summarised in a structured way, including a description of the organisation of energy management and an evaluation of its effectiveness. Topics discussed include energy and environmental policy, information systems, accountability and incentives, capital budgeting and investment criteria, purchasing and policy integration and awareness and organisational culture. Third, the case study results are analysed using the theoretical framework for the project and conclusions are drawn on which barriers are particularly important for the sector, together with the reasons for this. Finally, three types of initiatives which may help overcome the identified barriers are described, namely: measures which could be undertaken by the case study organisations; measures which could be undertaken by sector associations and other private sector bodies; and public policies to encourage energy efficiency.

Each chapter identifies similarities in the barriers faced by the case study organisations, together with important differences which result from variations in organisational size and structure, economic situation, incentive schemes, investment history and other factors. However, a theme common to all the country/sector studies is the importance of salary costs for energy management staff. These represent a very real hidden cost which in many cases will make the neglect of energy efficiency opportunities entirely rational.

In addition to the common elements described above, each chapter reports the results of some additional research. Chapter 3 explores the energy services market in Germany, including the nature and size of the current market, obstacles to the expansion of the market, and the scope for energy service companies to overcome barriers to energy efficiency within the German higher education sector. Chapters 4 and 5 include the results of the postal surveys on energy management practices conducted for the UK higher education and brewing sectors respectively, and these quantitative results are used when assessing the relative importance of different barriers in these sectors. Chapter 6 includes some quantitative analysis of the responses to the pre-interview questionnaires by the seven case studies within the Irish mechanical engineering sector.

In addition to the country/sector studies, Chapter 7 reports the results of additional research on the barriers to energy efficiency within the UK construction industry. This issue was found to be of particular importance within the higher education case studies, but the majority of the identified barriers appeared to lie outside the control of the host organisation. While these barriers are well known to construction industry specialists, they are relatively neglected in the energy economics literature. Furthermore, conventional policy measures such as building regulations appear to leave these barriers largely untouched. This chapter argues that the ongoing reforms to the UK construction industry have the potential to address many of the identified barriers, but this will only be achieved if sustainability and climate policy objectives are more directly integrated into the reform agenda

Finally, Chapter 8 summarises the overall results of the project and suggests answers to the original research questions. The chapter argues that both poles of the barriers debate are incorrect: technology optimists tend to underestimate the importance of hidden costs and other barriers, while economic pessimists tend to overestimate their importance and ignore policy options which may allow these barriers to be reduced. The chapter assesses the relative importance of different barriers within each of the case study sectors and demonstrates how 'hidden costs' and 'access to capital' were predominant themes within all of the organisations studied. It explores the determinants of these problems in more detail and argues that neither should be seen as beyond the appropriate scope of public policy. Instead, market and organisational failures should be seen as relative rather than absolute and contingent upon the particular form of economic organisation. While the neglect of no regrets opportunities is 'overdetermined' at present by the coexistence of multiple barriers to energy efficiency, these barriers may be lowered through either organisational or contractual measures or through particular types of public policy. This leads to an argument for a policy mix that can address the multiple sources of market and organisational failure and 'help organisations help themselves'.

NOTES

1. Both the EIA and IPSEP studies assumed uniform and consistent carbon pricing through the use of carbon taxes or carbon emissions trading. But the IPSEP study also assumed: a) the use of revenues from carbon taxes or allowance auctions to offset other forms of distortionary taxation; b) the use of the Kyoto mechanisms; c) the valuation of the health and air quality co-benefits associated with greenhouse gas emission reduction; d) the mitigation of non-CO_2 greenhouse gases, together with the use of carbon sinks and carbon sequestration; and e) the use of non-price regulatory policies to encourage the adoption of cost-effective energy technologies (IPSEP, 2001). While all of these policy options should feature in a cost-effective approach to mitigating climate change, it is the use of non-price regulatory policies which accounts for the majority of the difference between the two studies.
2. Bottom-up models are typically classified as either *optimisation* models, which use a linear programming framework to select a minimum cost option, or *simulation* models, which either use more straightforward techniques for investment appraisal or determine technology choices exogenously. Both are typically based upon databases of existing capital stock (for example, houses, appliances, boilers) and project how this stock ages and is replaced over time. Bottom-up models typically require extensive data sets and necessitate a difficult compromise between the realism of high levels of disaggregation and the constraints of data availability. In addition, optimisation models have the additional drawback of high

sensitivity – in that once a technology is found to be cost-effective, the model will tend to prefer it over all others.

3. Top-down models are typically classified as either computable general equilibrium (CGE) or econometric. CGE models operate by stimulating the operation of markets for factor inputs, products and foreign exchange, with equations specifying supply and demand behaviour across all markets which are solved for an equilibrium state. This produces a wealth of economy-wide information on market prices, sectoral output, investment, employment, foreign trade, energy consumption and emissions. However, this framework: a) assumes perfect competition and ignores market failures; b) assumes instantaneous adjustment and ignores adjustment costs; and c) lacks empirical validation, since the assumptions and relationships are not econometrically estimated. Econometric models, in contrast, are calibrated against time series data of relevant variables and can model the adjustment costs caused by economic changes, such as the imposition of a carbon tax. But the weakness of these models is that the relationships that held between variables in the past are assumed to remain fixed into the future. This makes such models inappropriate for analysing long-run changes or the economic effects of large changes in the demand/supply structure. Similarly, they have difficulty in investigating the impact of non-price regulatory policies (Zhang, 1995).
4. No regrets opportunities are defined by the IPCC as those options whose benefits, such as reduced energy costs and reduced emissions of local/regional pollutants, equal or exceed their costs to society, excluding the benefits of avoided climate change.
5. This was also a central claim of the 1997 Economists' Statement on Climate Change: '…Economic studies have found that there are many potential policies to reduce greenhouse-gas emissions for which the total benefits outweigh the total costs. For the United States in particular, sound economic analysis shows that there are policy options that would slow climate change without harming American living standards, and these measures may in fact improve U.S. productivity in the longer run.' (Redefining Progress, 1997). This statement attracted 2500 signatories, including eight Nobel Laureates.
6. The BARRIERS project (JOS3CT970022) was funded by the European Commission under the JOULE III programme. The full results of the project are available at http://www.sussex.ac.uk/ spru/environment/research/barriers.html.
7. Germany's target under the EU burden sharing agreement is to reduce greenhouse gas emissions by 21 per cent compared to 1990 levels, while Ireland's target is for a 13 per cent *increase* in emissions. But these differences in absolute numbers do not correspond to comparable differences in the costs of meeting each target.
8. The difference between case studies and survey methods is effectively captured by Sayer's (1992) distinction between intensive and extensive research designs.
9. This section is loosely based on the philosophy and methodology of critical realism. This was originally developed by Bhaskar (1975) and applied to economics by Lawson (1997). A very accessible introduction is provided by Sayer (1992).
10. Also, within social systems the mechanisms and processes which contribute to particular outcomes are likely to be unstable over time and space.

REFERENCES

Bhaskar, R. (1975), *A Realist Theory of Science*, Leeds Books, Leeds.

BRE (Building Research Establishment) (2000), *Getting Signed Up: Energy Services in the Public Sector*, Building Research Establishment, Watford.

Brown, M.A. (2001), 'Market failures and barriers as a basis for clean energy policies', *Energy Policy*, **29**(14), 1197–208.

Brown, M.A., M.D. Levine, W. Short and J.G. Koomey (2001), 'Scenarios for a clean energy future', *Energy Policy*, **29**(14), 1179–96.

Canes, M.E. (2002), *Economic Modelling of Climate Change Policy*, International Council for Capital Formation, Brussels.

De Almeida, E.L.F. (1998), 'Energy efficiency and the limits of market forces: The example of the electric motor market in France', *Energy Policy*, **26**(8), 643 –53.

DeCanio, S.J. (2003), *Economic Models of Climate Change: A Critique*, Palgrave Macmillan, New York.

DeCanio, S.J. (1994), 'Agency and control problems in US corporations: the case of energy efficient investment projects', *Journal of the Economics of Business*, **1**(1), 105–23.

Den Elzen, M.G.J. and A.P.G. De Moor (2003), 'Analysing the Kyoto Protocol under the Marrakesh Accords: economic efficiency and environmental effectiveness', *Ecological Economics*, **43**(2-3), 141–58.

Diesendorf, M. (1998), 'Australian economic models of greenhouse abatement', *Environmental Science and Policy*, **1**(1), 1–12.

EIA (Energy Information Administration) (1998), *Impacts of the Kyoto Protocol on US Energy Use and Economic Activity*, SR/OIAF/EIA-98-03 Energy Information Administration, Washington DC.

Energy Efficiency Office (1993), *Fuel Efficiency Booklet No. 4: Compressed Air and Energy Use,* ETSU, AEA Technology, Harwell.

Eto, J., S. Kito, L. Shown and R. Sonnenblick (2000), 'Where did the money go? The cost and performance of the largest commercial sector DSM programs', *The Energy Journal*, **21**(2), 23–49.

Golove, W.H and J.H. Eto (1996), *Market Barriers to Energy Efficiency: A Critical Reappraisal of the Rationale for Public Policies to Promote Energy Efficiency*, LBL-38059, Lawrence Berkeley Laboratory, University of California, Berkeley.

Grubb, M. and J. Kohler (2000), 'Induced technical change: evidence and implications for energy-environmental modelling and policy', Working Paper, Department of Applied Economics, Cambridge University.

Grubb, M., J. Edmonds, P. ten Brink and M. Morrison (1993), 'The costs of limiting fossil fuel CO_2 emissions: a survey and analysis', *Annual Review of Energy and Environment*, **18**, 397–478.

Helm, D. (2003), 'To good to be true', *The Utilities Journal*, **6** (March), 22–3.

Hensen, S.H. and J.C. Weisman (1998), *Performance Contracting: Expanding Horizons*, Fairmont Press, Lilburn.

ICCF (International Council for Capital Formation) (2002), *Kyoto Protocol and Beyond: Economic Impact on EU Countries*, ICCF, Brussels.

Interlaboratory Working Group (1997), *Scenarios of US Carbon Reductions: Potential Impact of Energy Efficient and Low Carbon Technologies by 2010 and Beyond*, Lawrence Berkeley National Laboratory, Berkeley, CA.

Interlaboratory Working Group (2000), *Scenarios for Clean Energy Future*, Lawrence Berkeley National Laboratory, Berkeley, CA (LBNL-44029).

IPCC (Intergovernmental Panel on Climate Change) (2001) *Climate Change 2001: Mitigation*, Cambridge University Press, Cambridge.

IPSEP (International Programme for Sustainable Energy Paths) (2001), *Cutting Carbon Emissions at a Profit: Opportunities for the US*, International Programme for Sustainable Energy Paths, El Cerrito, CA.

Jaffe, A.B., R. Stavins and R.G. Newell (1999), *Energy-Efficient Technologies and Climate Change Policies: Issues and Evidence*, Resources for the Future, Washington.

James, P. and P. Hopkinson (2002), *Service Innovation for Sustainability: A New Option for the UK Environmental Policy?*, Green Alliance, London.
Joskow, P. and D. Marron (1992), 'What does a negawatt really cost? Evidence from utility conservation programs', *Energy Journal*, **13**(4), 41–74.
Krause, F. (1996), 'The costs of mitigating carbon emissions: a review of methods and findings from European studies', *Energy Policy*, **24**(10-11), 899–915.
Laitner, J.A., S.J. DeCanio, J.G. Koomey and A.H. Sanstad (2003), 'Room for improvement: increasing the value of energy modelling for policy analysis', *Utilities Policy*, **11**(2), 87–94.
Lawson, T. (1997), *Economics and Reality*, Routledge, London.
Löschel, A (2002), 'Technological change in economic models of environmental policy: a survey', *Ecological Economics*, **43**(2-3), 105–26.
Lovins, A.B. and L.H. Lovins (1997), *Climate: Making Sense and Making Money*, Rocky Mountain Institute, Old Snowmass, Colorado.
Marsh, G., P. Taylor, D. Anderson, M. Leach and R. Gross (2003), *Options for a Low Carbon Future: Phase 2*, a Report for the Department of Trade and Industry, AEA Technology, Harwell.
Nordhaus, W.D. (2000), *Warming the World: Economic Models of Global Warming*, MIT Press, Cambridge.
Redefining Progress (1997), 'The Economists' Statement on Climate Change', available from: http://www.rprogress.org/publications/econstatement.html.
Romm, J.J. (1999), *Cool Companies: How the Best Businesses Boost Profit and Productivity by Cutting Greenhouse Gas Emissions*, Island Press, Washington DC.
Runde, J. (1998), 'Assessing causal economic explanations', *Oxford Economic Papers*, **50**(2), 151–72.
Sanstad, A.H., S.J. DeCanio, G.A. Boyd and J.G. Koomey (2001), 'Estimating bounds on the economy-wide effects of the CEF policy scenarios', *Energy Policy*, **29**(14), 1299–1311.
Sayer, (1992), *Method in Social Science: A Realist Approach*, Routledge, London and New York.
Sensenbrenner, F.J. (1998) *Opening Statement Before the Committee on Science – The Road from Kyoto: Part Four*, October 9, US House of Representatives, Washington DC.
Sorrell, S. (1999a), 'BARRIERS work package 2a: a taxonomy of barriers and suggested measures', a report to DG XII of the European Commission under the JOULE project 'Barriers to Energy Efficiency in Public and Private Organisations', SPRU – Science and Technology Policy Research, University Of Sussex, Brighton.
Sorrell, S. (1999b), 'BARRIERS work package 2b: interview protocol for the higher education sector', a report to DG XII of the European Commission under the JOULE project 'Barriers to Energy Efficiency in Public and Private Organisations', SPRU – Science and Technology Policy Research, University Of Sussex, Brighton.
Sorrell, S., J. Schleich, S. Scott, E. O'Malley, F. Trace, U. Boede, K. Ostertag, P. Radgen, (2000), 'Barriers to Energy Efficiency in Public and Private Organisations', final report to the European Commission, SPRU – Science and Technology Policy Research, University Of Sussex, Brighton.
Sutherland, R.J. (1996), 'The economics of energy conservation policy', *Energy Policy*, **24**(4), 361–70.

Weyant, J.P. and J. Hill (1999), 'Introduction and overview', in *The Costs of the Kyoto Protocol: A Model Evaluation*, special issue of *The Energy Journal*, vii–xliv.

Wilson, D. and J. Swisher (1993), 'Exploring the gap: top-down versus bottom-up analyses of the costs of mitigating global warming', *Energy Policy*, **21**(3), 249–65.

Yin, R.K. (1994), *Case Study Research: Design and Methods*, Sage Publications, London.

Zhang, X.Z. (1995), 'Economic approaches to cost estimates for limiting emissions of carbon dioxide', *International Journal of Environment and Pollution*, **5**(2/3), 194–203.

2. Understanding Barriers to Energy Efficiency

Steve Sorrell

'...the debate between "engineering" and "economic" approaches to energy efficiency in fact involves an underlying conflict over economic principles. To resolve this conflict, technology analysts must acknowledge that empirical findings are meaningful only when linked to a well articulated theoretical framework. Similarly, economists must recognise that theoretical assertions are meaningful only if they stand up to empirical scrutiny.' (Sanstad and Howarth, 1994a, p. 817)

1. INTRODUCTION

Disputes over the existence, size and policy relevance of an energy efficiency 'gap' have their roots in more fundamental disagreements over the role of the state in a market economy and the methodologies and concepts appropriate to studying economic behaviour. These differences reflect tensions between 'orthodox' economics and more recent developments in economic theory, as well as more fundamental conflicts between economics and other social sciences. But this dispute is not confined to academia. The existence or otherwise of an energy efficiency gap is central to the politics of climate change, and the nature and size of this gap is the subject of fierce debate between industrial and environmental interest groups. Despite innumerable studies from the mid-1970s onwards, the debate continues to be polarised between 'technological optimists' and 'economic pessimists', each of whom have their preferred concepts, methodologies and modelling tools. This creates a strong convergence of interest between those seeking to promote particular academic disciplines and those seeking convenient ways to justify their political positions.

This highly politicised context makes it all the more important to pay careful attention to the theoretical ideas that guide empirical research. The concepts, or hypotheses, need to be defined with care before their suitability for explaining the observed behaviour can be tested. This chapter provides a foundation for the empirical research by developing a taxonomy of six proposed barriers to energy efficiency and identifying the causal mechanisms which underlie these barriers. While the taxonomy is based upon categories that are widely used within the energy efficiency literature, the description and evaluation of the causal mechanisms draws upon more formal concepts, such as moral hazard and adverse selection. These more formal ideas are defined and explained prior to applying them to energy efficiency, and it is argued that several of these concepts are relevant to each of the proposed barriers.

The taxonomy synthesises ideas from three areas of economic theory: orthodox or neo-classical economics (including agency theory and the economics of information), transaction cost economics and behavioural economics. The boundaries between these approaches are blurred – for example, agency theory is often grouped with transaction cost economics (TCE) under the more general heading of the new institutional economics (Furubotn and Richter, 1997).[1] But to some extent, agency theory, TCE and behavioural economics can be seen as progressively greater departures from orthodox economic theory as it is portrayed in undergraduate textbooks. While agency theory and the economics of information are now very much part of the mainstream, transaction cost and behavioural approaches occupy a more ambiguous position.

The main argument of this chapter is that orthodox economic theory is necessary but insufficient to understand barriers to energy efficiency. Understanding may be enhanced by introducing concepts from more heterodox traditions, notably transaction costs and 'bounded rationality'. Not only do these concepts aid the understanding of barriers, they may also legitimate a wider scope for policy intervention.

The chapter is structured as follows. Section 2 introduces the barriers debate and the notion of an 'energy efficiency gap' and shows how orthodox economic theory can illuminate this debate. Section 3 reviews the core assumptions of orthodox theory and introduces the types of market failure that are relevant to the energy efficiency gap, focusing in particular on asymmetric information. Section 4 does the same for transaction cost and behavioural economics, identifying the nature and determinants of transaction costs, the importance of bounded rationality and the insights offered by social psychology. Section 5 draws upon ideas from each of these approaches to explore the taxonomy of barriers to energy efficiency and show how this encompasses barriers which are internal to organisations, as well as those within energy service markets. Each of the hypothesised barriers is illustrated with concrete examples from energy service markets or studies of organisational decision-making. Finally, section 6 summarises the taxonomy and discusses whether and how the proposed barriers may provide legitimate grounds for government intervention.

2. BARRIERS AND MARKET FAILURES

2.1 What is a Barrier?

In the present context, a barrier is a mechanism that inhibits a decision or behaviour that appears to be both energy efficient and economically efficient. In particular, barriers are claimed to prevent investment in cost-effective energy efficient technologies. To determine the existence of barriers it is necessary to ask (Weber, 1997):

- *What is the barrier?* For example, transaction costs, risk, lack of capital, split incentives, and so on.
- *Who or what is it an obstacle to?* For example, firms, public organisations, departments within organisations, individuals.
- *What does it prevent?* For example, purchase of more energy efficient equipment, retrofitting insulation to a building, establishing a monitoring and targeting scheme.

Hence for research to be fruitful it is necessary to:

1. Obtain a clear theoretical understanding of the nature of the proposed barrier. This involves identifying and abstracting the essential elements of a complex situation and requires assumptions about the nature of individual rationality (for example, utility maximising or satisficing).
2. Identify the relevant actor. The primary question is: why do organisations neglect energy efficiency? But to answer this, it is necessary to examine the role of individuals and departments within an organisation, including their contractual relationships with each other and with the outside world. In most cases a wide range of other actors will also be involved, including architects, designers, equipment suppliers, maintenance companies, public agencies and energy service companies. However, the primary focus of the present study is decision-making *within* energy-using organisations.
3. Identify the relevant decision or behaviour. Here, it is necessary to understand the general context of energy-related decision-making within an organisation, including the relevant structures, rules, obligations, incentives and routines. But within this, it is useful to focus on why a number of specific technologies have or have not been adopted. Hence, a

list of relevant energy efficient technologies has been developed for each of the case study sectors, based on those most widely recommended in energy efficiency publications.

This chapter is concerned primarily with (1) above – categorising the proposed barriers. The identification of the relevant actors, decisions and technologies forms part of the empirical research for each country and sector, which is described in chapters 3 to 7.

2.2 Barriers and Economic Theory

The term 'barrier', or 'market barrier', was introduced by researchers using engineering-economic models to study the technical and economic potential for energy efficiency. The observation that investments with a very high rate of return were being neglected led researchers to postulate that such investments were being inhibited by various barriers and that such barriers justified public intervention. The proposed barriers included split incentives, where the benefits of an investment did not accrue to the investor (the classic example is rental housing), and financing, where low income individuals were unable to gain access to capital for cost-effective energy efficiency investment. Many of these studies were conducted by energy efficiency practitioners rather than academic economists and were rather selective in their use of economic theory (Golove and Eto, 1996). For example, Hirst and Brown (1990) asserted that uncertainty in future fuel prices was a barrier, while Warren (1987) cited the use of stringent payback criteria for discretionary investment in industry. Studies such as these attracted criticism from orthodox economists who pointed out that the term 'barrier' was ambiguous and was not being used in a rigorous fashion (Sutherland, 1991). Many of the proposed barriers could be benign characteristics of well functioning markets (Sutherland, 1996, p. 363). For example, uncertainty in future fuel prices would lead risk-averse investors to substitute capital or labour for fuel, so should encourage energy efficiency, not act as a barrier (Sutherland, 1996, p. 363). Similarly, energy efficient investments may be associated with additional *hidden costs* such as management time or disruptions to production. Investors may be making a rational decision not to invest in the light of these costs.

Such criticisms have been very beneficial in clarifying ideas on barriers and relating these to traditional notions of market failures. The starting point is that public policy should aim to encourage economic efficiency and not energy efficiency alone (Jaffe and Stavins, 1994; Sutherland, 1996; Golove and Eto, 1996, p. 13). The basic theorems of orthodox economics demonstrate how, under a particular set of assumptions, competitive markets can maximise social welfare. But in some circumstances the requirements for efficient allocation of resources through well functioning markets can be violated. Perhaps the best-known example is where the production of a good leads to pollution which imposes costs upon individuals living near the manufacturing plant. If these external costs are not taken into account by the producer, overall economic efficiency will not be achieved. In these circumstances, the internalisation of the external costs by regulatory mechanisms such as a pollution tax may increase social welfare.

The pollution externality is one example of a *market failure* which is argued to provide a necessary condition for public intervention to improve economic efficiency (Weimer and Vining, 1999).[2] While the term barrier may refer to any factor which explains why technologies which appear cost-effective are not taken up, only a subset of these may correspond to recognised market failures. Hence, from the perspective of orthodox economics, only a subset of the identified barriers may justify policy intervention to improve economic efficiency (Jaffe and Stavins, 1994, p. 805). Furthermore, since market failures are pervasive, the mere existence of market failure is not sufficient to justify intervention. It is also necessary to demonstrate that the benefits arising from an intervention exceed the cost of implementation (Jaffe and Stavins, 1994, p. 808). The latter may be difficult to estimate but include government expenditure, investment by product manufacturers and loss of utility from changed product performance.

Jaffe and Stavins (1994) have used this framework to produce a helpful characterisation of the different types of barrier and of the orthodox rationale for public intervention. This is summarised in Table 2.1 (full explanations of the terms used in this table are provided in the subsequent sections). Note that while environmental externalities may justify public intervention, they do not explain the neglect of investments which appear cost-effective at current energy prices. Hence, while raising energy prices to reflect externalities is undoubtedly justified and should incentivise improvements in energy efficiency, this is not the primary focus of the present study. Instead, the aim is to identify the existence and relative importance of the postulated barriers in the first column of Table 2.1.

The Jaffe–Stavins framework provides a helpful characterisation of the potential for improving energy efficiency. It allows a distinction to be made between the 'economists' potential' achievable by removing market failures which explain the energy efficiency gap (upper left quadrant) and the 'technologists' potential' achievable by the additional removal of barriers which do not correspond to market failures (lower left quadrant). The technologists' potential is higher than the economists' potential, but achieving it would not contribute to economic efficiency as several of these barriers are not market failures. The 'theoretical social optimum' is higher still and allows for the additional removal of market failures which do not explain the energy efficiency gap, such as environmental externalities (upper right quadrant). But it is possible that only a portion of market failures can be removed at an acceptable cost. Hence, the 'true social optimum', which corresponds to the net effect of policies which pass a cost/benefit test, may be relatively small. In practice, however, the size of each potential is an empirical question and may be expected to vary with context.

Table 2.1 Barriers and orthodox market failures

	Explains efficiency gap	**Does not explain efficiency gap**
Barriers that are market failures	*Examples* • Public good attributes of information • Positive externalities of technology adoption • Asymmetric information in energy service markets – leading to problems of adverse selection, moral hazard and split incentives	*Examples* • Distortions in energy pricing (for example, departures from marginal cost pricing, cross subsidies, VAT differentials etc.) • Environmental externalities (for example, air quality, acid rain, climate change)
Barriers that are not market failures	*Examples* • Hidden costs (for example, disruptions to production) • Reduced product performance (for example, lower reliability) • Option value of delaying investment	

Note: The energy efficiency gap is defined as the existence of unexploited investment opportunities that appear worthwhile at current prices.

Source: Based on Jaffe and Stavins (1994).

Jaffe and Stavins also provide a useful discussion of the role of discount rates in the barriers debate (Jaffe and Stavins, 1994, p. 806). They note that the observation that organisations have high implicit discount rates when making energy efficiency decisions is merely a restatement of the existence of the efficiency gap. Estimates of implicit discount rates are derived through an examination of the decisions made and a calculation of the rates that would make those decisions optimal given estimates of capital costs and future energy savings. But simply observing that implicit discount rates are high says nothing about *why* they are high. This could either be through organisations applying normal discount rates in the context of hidden costs or market failure, or through organisations applying very high discount rates. Hence high discount rates are not a barrier, but a restatement of the phenomena to be explained (Jaffe and Stavins, 1994, p. 807). Similarly, a claim that the 'correct' discount rate is much lower than that observed amounts to assuming that all barriers are true market failures, while the claim that consumers are already using the correct discount rate amounts to assuming that there are no market failures (Sutherland, 1996). In neither case is the existence, operation or magnitude of the barriers demonstrated. The observation that discount rates are higher than those used in energy supply investment is open to the same criticism. There may be very good reasons why they are higher, related to the relative risk of investment. In sum, the debate over the correct discount rate to use is unproductive. The challenge is to demonstrate the operation of the proposed barriers empirically.

2.3 Limitations of the Market Failure Approach

The Jaffe–Stavins framework brings some much-needed clarity to the barriers debate and highlights the potential rationality of neglecting energy efficiency in a context of real but 'hidden' costs (hidden, that is, to the analyst but not to the investor). However, the framework has three important limitations. First, it treats market failures as absolute, in opposition to TCE which emphasises '…a broader category, that of transaction costs, which in general impede and in particular cases completely block the formation of markets…' (Arrow, 1969). Commentators using a TCE perspective regard energy efficiency as a co-ordination and incentive problem, rather than one of utility maximisation and emphasise that policy interventions and different institutional structures may lower transaction costs and provide net social benefits (Golove and Eto, 1996; Levine et al., 1994). This transaction cost perspective is explicitly rejected by Jaffe and Stavins, who appear to consider such costs as fixed and hence not an appropriate focus of public intervention (Jaffe and Stavins, 1994, p. 809).[3]

Second, the framework refers solely to *market* failures, and the examples used are drawn solely from energy service markets. For example, the problem of asymmetric information is illustrated with reference to housing markets. This is characteristic of the orthodox tradition, which has been reluctant to open up the black box of organisations and to analyse how individual organisations behave. Instead, the preferred approach has been to treat private firms as if they were unitary, profit maximising actors. The inadequacy of this approach has spurred the development of agency theory, TCE and other approaches which apply both traditional and new ideas to explain the structure and behaviour of organisations (Milgrom and Roberts, 1992; Douma and Schreuder, 1998).

Third, the framework utilises orthodox conceptions of economic rationality in which individuals and firms are assumed to solve complex optimisation problems, leading to outcomes which are consistent with the predictions of mathematical models. The lack of realism of this assumption has been a primary focus of criticism and has led to the development of alternative approaches such as evolutionary and behavioural economics (Hodgson, 1988; Nelson and Winter, 1982; Thaler, 1991).

These three themes are central to the discussion of barriers developed in the subsequent sections. The Jaffe–Stavins framework is taken as a starting point, but is extended by introducing transaction costs and more realistic representations of individual decision-making. The revised framework uses similar concepts to explain both *market failures* in energy service markets and *organisational failures* within organisations. The second category reflects the fact that many barriers result from organisational structures, procedures and routines, and from the incentives these provide to organisational members. The appropriate response may then be through managerial tools, such as incentive design, rather than public policy intervention. However, there may be circumstances where public intervention can help organisations overcome these internal barriers. Whether or not such failures justify either public intervention or a change in organisational arrangements will depend upon the relative costs and benefits of such interventions.

In addition to market and organisational failures, some barriers may explain the energy efficiency gap but not justify any action to overcome them. For example, if the installation of an energy efficient motor leads to interruptions to production, the savings in energy costs may be outweighed by the lost sales. Hence, a third category of *non-failure* is also appropriate, which includes hidden costs such as these. This leads to a threefold classification of barriers:

- *market failure*: where the barrier can be considered a market failure within an energy service market and may legitimate public policy intervention if the benefits exceed the costs;
- *organisational failure*: where the barrier results from organisational structure and policy and may be reduced through managerial tools such as task allocation and incentive design, and in some cases through public

policy measures which 'help organisations help themselves' (Golove and Eto, 1996, p. 30);[4] and

- *non-failure*: where the organisation is behaving rationally given the risk-adjusted rate of return on an investment in the existing context of energy, capital and unavoidable 'hidden' costs.

TCE argues that markets and organisations provide alternative means of organising economic activity and that the choice between them depends upon a number of factors including the relative magnitude of transaction costs (Williamson, 1979; 1985). So, for example, an organisation could choose to run non-core activities such as energy management in-house, or to outsource such activities to a specialised energy service company (ESCO). To the extent that both markets and organisations rely on explicit or implicit contracts, the categories of market and organisational failure could be brought together under a more general heading of *contract* failure (Furubotn and Richter, 1997, p. 131). The advantage of this perspective is that it recognises that transaction costs are contingent upon the particular form of economic organisation and that alternative forms of organisation may provide a means to overcome either market or organisational failures.

This classification of barriers is developed further in the subsequent sections and the distinction between hidden and transaction costs explained. The following two sections introduce orthodox, TCE and behavioural perspectives respectively, focusing on the definition of key concepts. The application of these concepts to energy efficiency is the subject of section 5.

3. ORTHODOX PERSPECTIVES

Orthodox, or neo-classical economic theory is characterised by the use of calculus, an emphasis on formal modelling and a focus on relative prices. Despite a century of increasingly elaborate development, the core assumptions remain largely unchanged. These include:

- *Methodological individualism*: The theory bases its explanations on the behaviour of individual members, rather than social structures. This is not applied consistently, however, since firms are traditionally treated as unitary actors.
- *Equilibria*: The primary interest is the allocation of scarce resources among alternative ends at a given moment in time. Evolution over time is traditionally ignored.
- *Self-interested maximising behaviour*: Primacy is given to utility and profit maximising in explaining observed economic behaviour and

organisation, rather than political power or social culture. Selfless motivations such as reciprocity and altruism are largely ignored.

- *Marginal trade-offs and optimisation*: Individuals are assumed to have stable and constant preferences which can be expressed by a utility function, while firms are assumed to have well-defined production or cost functions, exhibiting decreasing returns to scale. Individuals maximise utility subject to a budget constraint, while firms maximise profits subject to demand and cost conditions.
- *Rationality*: The previous assumption requires that individuals and firms are able to solve complex constrained optimisation problems instantaneously and at no cost.

The focus of orthodox theory is the allocation of scarce resources. Under certain conditions, a set of prices will arise in competitive markets that distributes inputs to firms and goods to individuals in such a way as to maximise social welfare.[5] Such a distribution is *Pareto efficient* in that noone can be made better off without making someone else worse off. The conditions required for Pareto efficiency include:

- a complete set of markets with well-defined property rights, such that all resources are fully allocated and privately owned;
- rational expectations by individuals and firms regarding prices and product characteristics; and
- competition between firms such that the action of an individual firm has no influence on product prices.

The violation of any of these conditions constitutes a market failure and hence a potential rationale for public intervention. Market failures are commonly grouped under four headings:

- incomplete property rights;
- positive and negative externalities;
- imperfect competition; and
- asymmetric information.

These categories overlap in that externalities can be understood as resulting from incomplete property rights (Bromley, 1991). Similarly, a commonly cited category of market failure, 'imperfect information', may result in part from incomplete property rights and the fact that information has the characteristics of a public good. Externalities and imperfect competition will not be considered further here since they are not directly relevant to the energy efficiency gap (that is, the upper left quadrant of Table 2.1). Instead, the focus will be on:

- imperfect information and information *as* a good; and
- asymmetric information and information *about* a good.

Consideration of asymmetric information also opens up some broader themes which are the primary focus of agency theory and the economics of information. Hence, this section will also consider three additional aspects of asymmetric information which are relevant to the energy efficiency gap, namely adverse selection, moral hazard and split incentives.

3.1 Imperfect Information and Information as a Public Good

Efficient markets require goods to be: *rivalrous*, where the consumption of a good by one person prevents its consumption by a second person; and *excludable*, where the owner of the good has exclusive control over its management and use, together with the benefits and cost of that use.[6] Market failure arises when goods are non-rivalrous in consumption, non-excludable in ownership, or both. For example, a lighthouse is non-rivalrous in consumption because ships may benefit from it without reducing the benefits to others. Similarly, ocean fisheries are (physically) non-excludable in consumption because fish can move freely between different regions. Such goods are commonly termed *public goods* and orthodox theory predicts that the absence of full property rights may lead to inefficiencies in either their supply or use. Table 2.2 provides a framework for classifying different types of goods.

Pure public goods are not depleted by an additional user and it is generally difficult or impossible to exclude people from their benefits. Because of the free-rider problem, they are unlikely to be provided by private firms and must instead be provided by a public agency. Examples of pure public goods include defence, public health, police and certain types of research. The relevance of this to the barriers debate is that *information* is considered to have aspects of a pure public good. That is, many people may 'consume' the same information without reducing its supply, while it may be difficult to exclude people from the benefits of information. Private firms may be unable to capture the full benefits of the sale or transfer of information because of the low cost of reproduction and distribution. Hence, they may be unwilling to supply information because they would be unable to capture all the benefits.

Table 2.2 A framework for categorising goods

	Rivalrous	**Non-rivalrous**
Excludable	*Private goods* • e.g. clothes, food, houses • Markets can be efficient	*Toll goods* • e.g. motorways, bridges • May be supplied by markets
Non-excludable	*Common property goods* • e.g. fisheries, common lands • Over-consumption and exhaustion of the resource is possible	*Pure public goods* • e.g. defence, public health, waste assimilation capacity of atmosphere • No private supply because exclusion is not possible

Source: Adapted from Weimer and Vining (1999, p. 80).

The public good aspect of information may be apparent in at least three circumstances. First, it may reduce the incentive to provide generic information about a particular class of good. More information may lead to an increase in consumer demand, but this increased demand may expand the whole market rather than advancing the share of the supplier who disclosed the information. The net result may be too little information about the product class (for example, energy efficient motors) and too much information about the relative advantages of a particular model (Office of Fair Trading, 1997).

Second, it may inhibit the development of market intermediaries, such as agents and certification services, who provide an alternative source of such generic information. If the good is relatively homogeneous, a consumer could easily pass along any information that the agent provides to other prospective purchasers, thereby reducing the demand for agency services (Weimer and Vining, 1999, p. 74).

Third, it may create a disincentive to adopt new technologies, since the information created by this adoption (performance, risk, reliability and so on) has the characteristics of a public good. If this information is available to competitors it will reduce the risk associated with the subsequent adoption of the technology, but the value inherent in this reduced risk will not be captured by its creator (Golove and Eto, 1996, p. 19). This is a common rationale for government-funded demonstration projects for new technologies.

The public good nature of information may contribute to *imperfect information* regarding: the price of a product and of other products (complements and substitutes); the terms of trade (for example, location of supplier, date of delivery); and/or the quality of a product along a range of dimensions, relative to substitutes (Office of Fair Trading, 1997). The last of these is particularly relevant to energy efficiency, since this is normally just one attribute of a good whose utility is assessed along a range of dimensions (of which energy efficiency is rarely the most important).

If a consumer does not have all the relevant information about a good, they may suffer a loss of utility – perhaps by purchasing more or less of the good than they would under perfect information, or by purchasing a product with different attributes from those they would prefer. Experience suggests that this condition of imperfect information is a pervasive feature of most types of transaction, although the extent of imperfection varies widely with the nature of the good and the particular aspect of information that is being considered. Such imperfections may lead to a loss of utility which, given the public good aspects of information, may justify policy intervention to improve information in some circumstances – for example by labelling schemes, disclosure requirements or government testing of products.

In some cases, information may be under-supplied because firms cannot appropriate the full benefits. But in other cases, consumers may lack information because it is *costly* to acquire (Mohlo, 1998). The 'economics of information' was initiated by Stigler (1961) who assumed that consumers engage in search processes to narrow the gap between their expectations regarding different product attributes and the actual state of affairs. In a simple maximising model, the search process should proceed until the expected marginal value of an additional unit of information equals the anticipated marginal cost of searching. But some information may be uncertain or impossible to obtain. For example, the optimal decision may depend upon future events or factors which are outside the control of both the consumer and producer. In a world of uncertainty and positive information costs, the outcome of a rational (that is, optimising) decision-making process will depart from that in a world of perfect information.

Given these features of the consumer decision-making environment, the mere existence of imperfect information does not justify policy intervention. Instead, it is necessary to demonstrate either that the public good aspect of information is particularly significant, or that the cost of information acquisition inhibits transactions that would increase net social welfare. In a great many cases, the relevant information may be available to other parties to a transaction (for example, the seller), but the transfer of this information may be costly, or the other party may have no incentive to do so. This condition of *asymmetric information* is widely applicable and can provide some valuable insights into the operation of markets, the behaviour of economic actors and the structure of individual organisations. The following sections explore these implications in more detail.

3.2 Asymmetric Information and Information about a Good

Asymmetric information exists where the producer or supplier of a good holds relevant information, but is unable or unwilling to transfer this information to prospective buyers.[7] The extent to which asymmetric information leads to market failure will depend upon the nature of the good. Economists commonly classify goods into three categories:

- *search goods*: where a consumer can determine characteristics with certainty prior to purchase;
- *experience goods*: where consumers can only determine characteristics after purchase; and
- *credence goods*: where it is difficult for consumers to determine quality even after they have begun consumption.

In addition the importance of information asymmetry is influenced by the *variance* in quality of a good, the *frequency* of purchase relative to changes in underlying characteristics, and the *search costs* incurred in obtaining relevant information (Weimer and Vining, 1999, p. 110).

With search goods, consumers incur costs in inspecting different units of a good and different supply sources (for example, visiting several shops). Information asymmetry is only likely to be a problem when search costs are high relative to the purchase price, the good is heterogeneous in price and quality and the frequency of purchase is relatively low. In these circumstances, consumers may potentially forego favourable purchases. But even here, market mechanisms such as advertising and branding of products can reduce such inefficiencies (Weimer and Vining, 1999, p. 111).

With experience goods, consumers can only determine the quality of a good through consumption. This feature makes experience goods more prone to market failure. As with search goods, the potential for inefficiency is greater when a good is heterogeneous in quality or when the frequency of purchase is low relative to changes in the quality of the good. However, unlike with search goods, inefficiency may result in consumers regretting the purchase of a good, or even being harmed by its consumption. Such inefficiency may be mitigated through advertising and branding of goods, by the development of certification services which guarantee minimum quality standards, or by market intermediaries that sell advice about the quality of expensive, infrequently purchased, heterogeneous goods such as houses. Market failure may still occur, however, when quality is heterogeneous and unstable over time, branding is ineffective for various reasons, the growth of secondary markets is inhibited by the public good characteristics of information, or where market intermediaries are expensive relative to the price of the good (Weimer and Vining, 1999, p. 112). In these circumstances, there may be a case for public intervention.

With credence goods, consumers may not be able to identify the true quality of a good even after consumption. For example, certain types of drugs may lead to long-term health effects whose cause may be unclear to the sufferer many years after the original exposure. This type of good is particularly prone to market failure and in some cases this may have serious consequences. As a result, certain types of credence goods – for example, medical pharmaceuticals – are a common target of government regulation.

Search, experience and credence goods may therefore be placed on a spectrum, with increasing risk of market failure from information asymmetry. In some cases, secondary market mechanisms such as certification schemes may correct such failures, but these will not operate effectively in all circumstances and may be inhibited by public good problems. Hence, there may be circumstances where public policy can cost-effectively overcome such market failures.

3.3 Adverse Selection

If the producer or supplier of a good has difficulty in communicating information to a potential buyer, the latter will be unable to observe the superior quality of a product and will tend to select goods on the basis of visible aspects such as price. Although buyers may be prepared to pay a price premium for high quality products, this preference will not be exercised if product quality cannot be observed.

In other cases, it may be possible for the seller to inform potential buyers about product quality but s/he may not have the incentive to do so. In this case, the condition of asymmetric information provides an opportunity for the seller to act *opportunistically* to the detriment of the buyer. The assumption of opportunism departs somewhat from the more innocent form of self-interest seeking that is traditionally assumed in orthodox theory, but it is a core assumption of both agency theory and TCE. Opportunism is defined by Williamson as:

> '… self interest seeking with guile. This includes but is scarcely limited to more blatant forms, such as lying, stealing and cheating…more generally, opportunism refers to the incomplete or distorted disclosure of information, especially to calculated efforts to mislead, distort, disguise, obfuscate, or otherwise confuse.' (Williamson, 1985, p. 47).[8]

It is possible for a combination of asymmetric information and opportunism to lead to *adverse selection*, where the price charged for a product affects the selection of products that are bought or sold (Milgrom and Roberts, 1992, p. 149). The classic demonstration of this was by Akerlof (1970) who developed a model of the second-hand car market. The quality of a second-hand car is largely private information to the seller, since the buyer has only imperfect means to assess the quality of a car prior to purchase (that is, second-hand cars are experience goods). In this situation, some sellers will opportunistically supply poor quality cars. The more poor quality cars there are on the second-hand market, the lower the average quality and the lower the price. This will discourage owners of quality cars from putting them up for sale, thereby driving the average quality and price still lower. This sort of process could reach an equilibrium in which only the poorest quality cars are traded – in other words, poor quality products could drive high quality products out of the market.[9]

The problem of adverse selection is not confined to the purchase of goods. Similar examples can be found in loan contracts,[10] lease contracts, employment contracts and insurance contracts (Furubotn and Richter, 1997, pp. 132–38). A common feature of all these situations is that asymmetric information provides an incentive for one party to act opportunistically *prior* to entering into a contract. In other words, adverse selection result from pre-contractual opportunism (Mohlo, 1998, p. 8).

The problem of adverse selection may be overcome through market arrangements such as screening (Stiglitz, 1975) and signalling (Spence, 1973).[11] Signalling is the attempt to communicate private information in a credible way, for example by offering free maintenance or warranty in the car example, or by acquiring educational qualifications to signal productivity in the labour market example. Screening is the design of products or contracts to attract only desirable groups, for example by offering jobs with incentive pay schemes to attract the most productive workers (Milgrom and Roberts, 1992, pp. 154–9). But as with asymmetric information regarding product characteristics, these mechanisms may not always be effective and there may be circumstances where public intervention is appropriate.

3.4 Moral Hazard

In contrast to adverse selection, the problem of *moral hazard* results from post-contractual opportunism, in that asymmetric information provides actors with the opportunity to cheat *after* a contract has been signed (Mohlo, 1998, p. 8).

The term moral hazard originated in the insurance industry (Pindyck and Rubenfeld, 1998, p. 629). Insurance companies lack the ability to monitor the actions of policyholders after they have issued a policy. But the holder may act in a way to make the insured loss more likely to occur. The term is now used more widely to describe any behaviour under a contract that is inefficient (Milgrom and Roberts, 1992, p. 195). It refers to situations where the actions of one party are unobservable to a second party. Since the interests of the two parties may differ, this creates an incentive for one party to act in an opportunistic manner to the detriment of the other. For example, an employee may engage in 'shirking', because the employer is unable directly to monitor the amount and quality of the employee's effort (Milgrom and Roberts, 1992, p. 179). In general, moral hazard results when there is a divergence of interests between the parties to a contract, combined with difficulties in monitoring performance and enforcing contract terms.

Moral hazard is usually discussed in the context of *principal-agent* relationships (Holmstrom, 1979). Here, the agent is the party who acts (for example, a worker) and the principal is the party whom the action affects (for example, an employer). The principal's problem is to ensure that the agent acts to her benefit, but she lacks complete information – for example, she cannot evaluate how hard the agent has worked or whether she has been honest. The literature on 'agency theory' began by examining agency relationships between managers of companies (the agent) and shareholders (the principal), but has been extended to agency relationships in a wide range of contexts (Milgrom and Roberts, 1992; Prendergast, 1999). For example, in a sales contract, the supplier may fail to provide adequate servicing or may charge excessively for replacement parts. In a rental contract, the tenant may fail to maintain the property adequately while the landlord may be unable to observe the resulting deterioration. In a construction contract, subcontractors may be able to substitute cheaper equipment, which is less durable or reliable. These examples demonstrate that principal-agent relationships pervade both markets and organisations and that individuals are typically engaged in multiple relationships in which they act as both principals and agents.

There are a number of means available to control moral hazard. These include reducing information asymmetry by increasing the resources devoted to monitoring and verification; introducing incentive contracts that pay for performance along a measurable dimension; posting a 'bond' that can be forfeited if an agent is caught cheating; and harmonising incentives through vertical integration or long-term contracts (Milgrom and Roberts, 1992, p. 195). The feasibility and cost of such measures will vary with context, and if the costs are excessive they will not be used. As with adverse selection, the scope for moral hazard may be reduced through changes in contract terms and organisational arrangements, as well as through public policy measures.

As before, the costs and benefits of such measures would need to be carefully assessed.

3.5 Split Incentives

A final example of potential market failure resulting from asymmetric information is split incentives. This is commonly cited in the energy efficiency literature, particularly in relation to rental housing. The landlord of a building may be unwilling to retrofit an apartment to reduce energy use since the resulting savings would be realised by the tenant. But at the same time tenants will be unwilling to retrofit since they may move out before benefiting from the cost savings. Since neither party is able to appropriate the benefits of the investment, highly cost-effective opportunities may be ignored.

The analogy to the previous examples is that one party may have relevant information on the costs and benefits of an energy efficiency investment, but may find this difficult to convey to the other party (Jaffe and Stavins, 1994, p. 805). If there were no information problems, landlords and tenants would be able to enter into contracts to share the costs and benefits of the investment. For example, landlords could invest in energy efficiency measures and recover the value of that investment through higher rents. In practice, however, the gains that would be achievable through such arrangements are likely to be outweighed by the transaction costs involved. The arrangements that would be needed to make rents reflect the capitalised value of energy savings could be complex. As a consequence, the investments are likely to be foregone despite the potential advantages to both parties.

As with other forms of asymmetric information, there may be circumstances where policy intervention can reduce this market failure. For example, mandatory requirements to conduct energy audits on commercial buildings, coupled with standardised labelling schemes for energy performance, may give confidence to landlords that energy efficiency improvements will be reflected in rental values.

3.6 Summary of Orthodox Perspectives

The orthodox perspective starts with a highly restrictive set of assumptions regarding the operation of idealised markets. These include perfect information, zero transaction costs and maximising behaviour by individuals and firms. Economists then identify a number of departures from these assumptions, termed market failures, which may in some circumstances legitimate policy intervention to improve economic efficiency. These include incomplete property rights, externalities, imperfect competition and asymmetric information.

In practice, the orthodox assumptions bear little resemblance to real-world economic decision-making, which is characterised by imperfect information, positive transaction costs and substantial departures from maximising behaviour. Most real-world economic decisions are made with imperfect information regarding product prices, the terms of trade and (in particular) product quality. While imperfect information may partly be understood as a consequence of the public good nature of information, this somewhat abstract argument does not take us very far. Instead, what is more relevant is the cost of obtaining information, the asymmetric distribution of this information between different parties to a contract, and the tendency of actors to use this information opportunistically. By incorporating these features into the orthodox framework, it is possible to obtain some valuable insights into the informational problems associated with different types of good, the origin, nature and consequences of adverse selection and moral hazard, and the problem of split incentives. It will be argued below that these concepts are highly relevant to the barriers debate, and allow the empirical observations of energy researchers to be interpreted within a rigorous theoretical framework.

However, while this elaboration of orthodox theory has much to offer, it still has some limitations. For example, the core assumption of maximising behaviour remains largely unchallenged, despite its lack of realism. The optimising approach implies, for example, that agents can foresee what additional information further search would bring, what it would cost and what opportunities would be foregone by that search. Similarly, agency theory lacks consistency in that it frequently assumes perfect information in some areas and imperfect information in others (Furubotn and Richter, 1997). Much of the agency literature employs complex and highly formalised mathematical models whose utility in explaining real-world organisational arrangements must be questioned. Transaction cost economics and behavioural economics seek to overcome these weaknesses, principally by incorporating more realistic models of human decision-making. Together, they may provide a more comprehensive account of economic organisation, as well as contributing additional insights into the barriers debate.

4. TRANSACTION COST AND BEHAVIOURAL PERSPECTIVES

4.1 Transaction Cost Economics

Transaction cost economics (TCE) builds upon and extends the concepts introduced in the previous sections (Williamson, 1985; 1989). In common with orthodox theory, it bases explanations on the behaviour of individuals rather than social structures and assumes these individuals to be *intendedly* rational, in that they seek out opportunities to improve economic efficiency. In common with agency theory, TCE attaches particular importance to asymmetric information and opportunism. But TCE extends the orthodox/agency framework by first, introducing the behavioural assumption of *bounded rationality*; second, focusing on the nature of different *transactions* and the costs and risks associated with those transactions; and third, explaining why particular types of transaction are associated with particular types of *governance structure*. The latter is assumed to include both markets and hierarchical organisations, together with a wide variety of intermediate forms.

The concept of bounded rationality was first introduced by Simon (1957; 1959), who argued that behavioural models should be descriptively accurate. Orthodox economics fails in this respect, since it assumes that individuals make decisions in the manner prescribed by formal optimisation models – or that their choices are consistent with the predictions of such models. In contrast, bounded rationality assumes that people make decisions subject to constraints on their time, attention, resources and ability to process information – and hence that their choices are likely to differ from the predictions of optimisation models. Boundedly rational individuals will economise on scarce cognitive resources by utilising routines and rules of thumb and will tend to make satisfactory decisions rather than expend time and effort searching for the optimum decision – a process that Simon termed *satisficing*.[12]

TCE improves upon the orthodox approach by recognising the central importance of bounded rationality, but stops short of exploring the full implications of this for individual decision-making (Foss, 2003). Instead, TCE employs bounded rationality as a background assumption that has two important consequences:

- *Contracts are incomplete*: Since agents do not have the capacity to foresee every contingency that might arise or every consequence of any action they may take, any contracts they engage in will necessarily be incomplete in that they will not specify the actions to be taken in all circumstances.[13]

Incomplete contracts must specify the procedures for dealing with such contingencies, while at the same time safeguarding against the risk of opportunistic behaviour by either party (Williamson, 1985).

- *Transaction costs are unavoidable*: If individuals are boundedly rational, they *must* incur transaction costs in decision-making – regardless of the relative availability of information. Hence, transaction costs are encountered universally, within both markets and organisations, because of the character of the individuals who make decisions (Furubotn and Richter, 1997, p. 39).

Transaction costs are assumed to include the legal, administrative, information gathering and other costs associated with searching for partners, negotiating and writing contracts, monitoring performance and enforcing promises. The search costs analysed in the economics of information form an important component of transaction costs. But the term is used in a broader sense to include all the costs associated with establishing, maintaining and transferring property rights – where the latter are defined as all types of public and private rules and organisational arrangements (Allen, 1996). So there will be transaction costs within organisations, associated with decision-making, monitoring performance, transferring information and so on, which parallel the search and information costs within markets (Table 2.3). TCE argues that transaction costs provide the primary explanation for the particular form taken by both large-scale economic institutions and specific contractual relationships (Williamson, 1985).[14]

In practice, transaction costs are extremely difficult to measure and determining their absolute value in any particular instance is likely to be problematic. TCE chooses instead to focus upon the determinants of transaction costs, their relative size in different circumstances and how they can be minimised by the choice of an appropriate governance structure (for example, spot market or long-term contract). The basic assumption is that market actors choose governance structures to minimise transaction costs – or more specifically '…to economise on bounded rationality while at the same time safeguarding against the hazards of opportunism' (Williamson, 1985, p. 32). From a TCE perspective, market failures and organisational failures are two sides of the same coin: each results from excessive transaction costs and each may potentially be mitigated by substituting one form of organisation for another, or for an intermediate form of organisation such as a joint venture. The TCE literature is therefore less concerned with identifying appropriate circumstances for public intervention than with using transaction costs to explain the observed *choice* of market, organisational and contractual forms and the specific *design* of individual contractual arrangements.

TCE uses four variables to characterise transactions:

- the *uncertainty* associated with the transaction (resulting both from exogenous factors and the potential to behave opportunistically);
- the *frequency* with which the transaction is repeated;
- the *complexity* of the transaction; and
- the extent to which parties to a contract must invest in assets which are *specific* to that relationship (Williamson, 1989).

The last of these factors is given particular emphasis and refers to 'the degree to which an asset can be re-deployed to alternative uses and by alternative users without sacrifice of productive value' (Williamson, 1989, p. 142). An example of high asset specificity is a coal-fired power station that locates close to a separately owned coal mine. Once established, the power station is vulnerable to 'hold-up' by the mine owners, in that the station has limited bargaining power should the mine owners demand a higher price for the coal (Joskow, 1985). To protect such assets, simple market relationships are likely to be replaced by alternative forms of governance that provide safeguards to mitigate such risks, such as long-term contracts or unified ownership.

TCE uses uncertainty, frequency, complexity and 'asset specificity' to explain the choice of governance structure for different types of transaction. Governance structures are located along a spectrum, with spot markets at one pole and hierarchical organisations at the other (Williamson, 1985). Market structures provide powerful incentives for exploiting profit opportunities and allow quick adaptation to changing circumstances, but they expose parties to the risk of opportunistic behaviour when investment in specific assets is required. In contrast, more hierarchical structures reduce the scope for opportunistic behaviour but provide managers with weaker incentives to maximise profits and lead to additional bureaucratic costs. In general, hierarchy is more likely to be chosen when transactions are complex and frequent, uncertainty is present and investment in specific assets is required, while spot markets will be chosen when the opposite is the case.

For our purposes, TCE provides a comprehensive framework through which to understand economic organisation in general and barriers to energy efficiency in particular. It recognises that all types of transaction have associated costs which are contingent on the particular market, organisational and contractual arrangements. This implies that changes to these arrangements may lower transaction costs and thereby improve economic efficiency. Such changes may be initiated by private actors, for example by introducing organisational incentive schemes or contracting out non-core services, or they may be facilitated by public policy, such as when standardised labelling schemes reduce the search costs for energy efficient motors. In this way, transaction costs can provide an overarching category that to a large extent subsumes the orthodox market failures resulting from asymmetric information.

Table 2.3 Types of transaction costs

Type		**Examples**
Market (external)	Search and information costs	Searching for parties with whom to deal; communicating; gathering information about price and quality. These costs are the subject of the economics of information.
	Bargaining and decision costs	Bargaining and negotiating costs; time and legal advice; costs of making any information gathered usable; compensation paid to advisers; cost of reaching decisions within groups.
	Supervision and enforcement costs	Monitoring agreed upon delivery times; measuring product quality and amounts; measuring the valuable attributes of what is being exchanged; protecting rights; enforcing contractual provisions.
Organisational (internal)	Establishing organisations	Costs of setting up, maintaining or changing and organisational design, including incentive design, information technology, public relations, lobbying, etc.
	Running organisations	Costs of decision-making, monitoring the execution of orders, measuring the performance of workers, agency costs, costs of information management etc. Costs associated with physical transfer of goods and services across an interface, including idle time, intra-firm transport, logistics, quality control, etc.

Note: Here, 'market' transaction costs include those incurred by organisations when engaging in external transactions, while 'organisational' transaction costs relate solely to internal transactions.

Source: Based on Furubotn and Richter (1997, pp. 43–7).

TCE assumes that the observed market and organisational forms can be explained in terms of minimising transaction costs in the context of competitive market pressures. This involves an appeal to evolutionary selection, but only in a weak, satisficing sense: '...the *fitter* survive, but there is no reason to suppose that they are fittest in any absolute sense' (Simon, 1983, p. 69). Hence, inefficiencies may exist and these may in some circumstances be reduced. The scope for this can only be evaluated on a case by case basis, but such assessments may be facilitated by comparative studies or by examination of best practice. While transaction costs may be difficult to quantify, it should be possible to identify their determinants and assess their *relative* magnitude under different policy or organisational arrangements.

4.2 Behavioural Economics

TCE extends the scope of the orthodox model, adds greater realism to the behavioural assumptions and has been applied in a large number of empirical studies (Shelanski and Klein, 1995). But despite introducing a more realistic model of individual decision-making, TCE fails to 'flesh-out' the full implications of bounded rationality. Foss (2003) expresses the problem as follows:

> '...The bounded rationality research effort may be understood as an attempt to elaborate and examine the insights that (1) the human capacity to process information is quite limited, (2) humans tried to economise on cognitive effort by relying on shortcuts, and (3) because of (1) and (2), as well as other factors...human cognition and judgement is subject to a wide range of biases and errors. Economists of organisation have taken the interest in (1), have been less occupied with (2), and have almost entirely neglected (3).' (Foss, 2003, p. 259)

To explore the implications of (3), it is necessary to take account of the psychology-based research on *biases* to human cognition and judgement (Conlisk, 1996; Camerer, 1998; Rabin, 1998; Kahneman and Tversky, 2000). This research identifies how human decision-making systematically deviates from the expected utility model assumed in orthodox economics and shows how these deviations are both regular and predictable. The mental shortcuts that individuals use to economise on bounded rationality do not lead solely to 'limitedly rational' decisions but to systematically biased and erroneous decisions (Piattelli-Palmarini, 1994).

The recognition of such biases has led to the development of *behavioural economics*, whose main proponent, Daniel Kahneman, was awarded the Nobel Prize in 2002. Behavioural economics argues that the biases within human decision-making need to be taken seriously if a fully explanatory account of economic organisation and behaviour is to be provided, and if the predictive capability of economic models is to be improved. Kahneman and others have used experimental tests of decision-making under a variety of conditions to demonstrate that such biases are universal, predictable and largely unaffected by either monetary incentives or learning (Kahneman and Tversky, 2000). Kahneman and Tversky have developed 'prospect theory' to explain such biases and to provide a descriptively accurate alternative to the assumptions of orthodox economics (Kahneman and Tversky, 1979). But, to date, the implications of prospect theory have received only limited attention outside behavioural and experimental economics (Thaler, 2000; Rabin, 1998).[15] And given the constraints of case study research, the implications of this work can only be taken into account in a very limited way in the present study. But two important implications may usefully be highlighted: the certainty effect and loss aversion.

Kahneman and Tversky (2000) have demonstrated that outcomes received with certainty are weighted more heavily than uncertain outcomes. For example, most people would prefer £500 with certainty to a 50 per cent chance of winning £1000. But the orthodox model maintains that these two options are equivalent. Moreover, individuals have been shown to under-weight outcomes with a low probability and over-weight those with a high probability, leading to risk aversion with respect to gains and risk seeking with respect to losses. For example, subjects are willing to accept an 80 per cent risk of losing £4000 in preference to a sure loss of £3000, but they are more likely to choose a sure gain of £3000 than an 80 per cent chance of winning £4000. These preferences are relevant to investment appraisal, since they imply that business decisions may deviate from the predictions of orthodox economics.

Loss aversion appears in a variety of forms, including the 'endowment effect' first noticed by Thaler (1980). In one experiment, subjects were asked to imagine that they held a particular job and had the option of moving to a different job which was better in one respect and worse in another. Most subjects who were assigned to the first job did not wish to move to the second, and most subjects who were assigned to the second did not wish to move to the first (Kahneman et al., 1991). Thaler explains this by arguing that individuals generally require more money in compensation to give up something than they would pay to obtain it. Hence, opportunity costs are not treated as equal to out-of-pocket costs, and foregone gains are considered to be less painful than perceived losses. Similar observations are made by Samuelson and Zeckhauser (1988), who use the term 'status quo bias'. Many decisions take the form of a choice between retaining the status quo and accepting an alternative which is advantageous in some respects and disadvantageous in others. The advantages of alternative options will then be evaluated as gains and the disadvantages as losses. Since individuals place greater weight on losses than on gains, the decision is biased in favour of retaining the status quo.

Kahneman and Tversky (1979) explain these observations by suggesting that the outcome of a risky decision is evaluated by value function that has three characteristics. First, value is defined in terms of changes from a reference point, rather than absolute states of wealth or welfare. Second, the marginal value of both gains and losses decreases with their size (for example, an increase in a prize from £10 to £20 means more than an increase in a prize from £100 to £110). Third, the value function is steeper in the negative than in the positive domain, which means that losses loom larger than corresponding gains. Together with associated assumptions about the certainty effect, this framework can explain a wide range of observations of decision-making under risk. One important consequence is that a degree of *inertia* is introduced into both consumer and organisational choice, since there is a preference for the status quo (Thaler, 1991). A second is that that individuals can no longer be said to have a well-defined preference function, because choices will depend strongly on how a decision is framed (for example, as a gain or a loss) (Kahneman and Tversky, 1979).[16]

Since these observations run counter to the assumptions of orthodox economics, the suitability of the latter for explaining observed economic behaviour must be questioned. And since the objective of the present study is to *explain* the observed neglect of energy efficiency opportunities, the insights available from behavioural economics should be given some consideration.

4.3 Summary of Transaction Cost and Behavioural Perspectives

Taken together, the transaction cost and behavioural perspectives can provide a richer and more realistic account of economic behaviour and organisation than is available from within the orthodox model. TCE builds upon and extends a number of themes within agency theory to provide a unified approach to both market and organisational failures which emphasises that these are relative rather than absolute. Behavioural economics goes further by incorporating a more realistic account of individual decision-making which includes loss aversion and the over-weighting of certainty. But while both provide greater realism, they make it more difficult to develop formal models or to draw clear policy conclusions. In particular, the insights offered by Jaffe and Stavins into the barriers debate become undermined once transaction costs blur the simple distinction between market failures and non-failures. In a similar manner, the distinction between market failure and organisational failure becomes less robust when each may potentially be remedied by substituting one form of organisation for another.

Transaction costs are notoriously difficult to measure, with the result that '...there is a suspicion that almost anything can be rationalised by involving suitably specified transaction costs' (Williamson, 1979, p. 233). Instead of measuring transaction costs, the more usual approach is to develop proxy measures for uncertainty, complexity, frequency and asset specificity, and to predict the resulting governance structure (Shelanski and Klein, 1995). But while this allows very general hypotheses to be tested, such as the choice between debt and equity financing for particular type of investment, these are far from precise (Williamson, 1988). Behavioural economics poses even greater problems as the models developed from small-scale experiments are difficult to specify and have yet to be applied to larger-scale problems of organisational structure and behaviour (Foss, 2003). While both approaches have stimulated a great deal of empirical research, their application to energy efficiency remains in its infancy (Haddad et al., 1998).

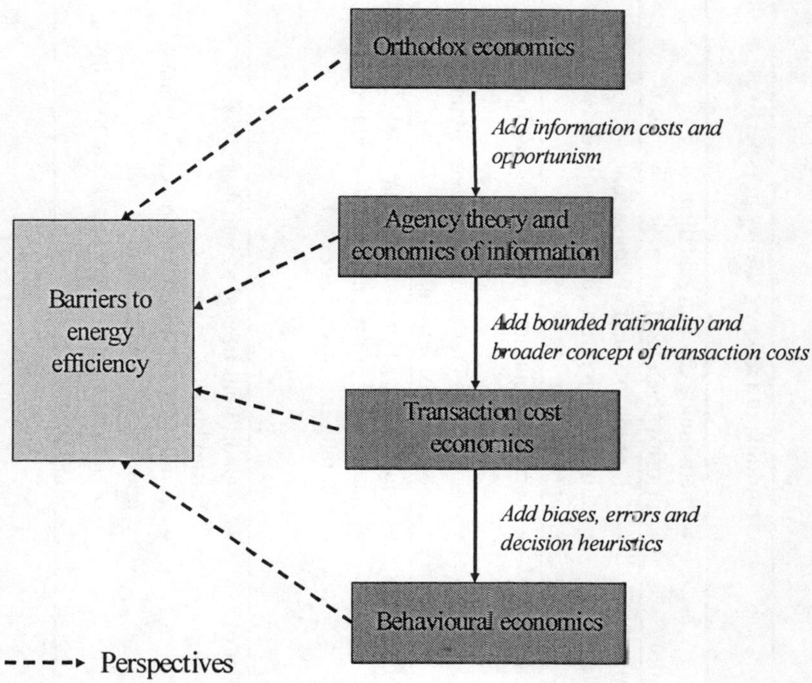

Figure 2.1 Extending the orthodox model

Table 2.4 Economic perspectives on the barriers debate

	Orthodox and agency theory perspectives	**Transaction cost and behavioural perspectives**
Basis of explanation	• methodological individualism • priority given to efficiency	• methodological individualism • priority given to efficiency
Rationality assumptions	• full rationality and optimisation • simple self interest seeking (orthodox) • opportunism (agency)	• bounded rationality and satisficing • opportunism • biases and cognitive illusions
Transaction costs	• complete contracts and zero transaction costs (orthodox) • search costs in markets analysed within the economics of information	• incomplete contracts and positive transaction costs • transaction costs inherent in all types of decisions and contractual relations
Focus of analysis	• equilibrium, efficiency and marginal trade-offs • maximising utility and profit (orthodox) • aligning incentives (agency)	• attributes of transactions and choice of governance structures • economising on bounded rationality while safeguarding against the hazard of opportunism
Treatment of organisations	• organisations treated as unitary actors (orthodox) • asymmetric information allows a common approach to a wide range of market, organisational and contractual relationships (agency)	• organisations treated as a nexus of contracts • transaction costs allow a common approach to a wide range of market, organisational and contractual relationships

Understanding of market failures	• market failures are absolute • result from incomplete property rights, externalities, imperfect competition and asymmetric information	• market failures are relative • result from excessive transaction costs • cognitive biases may be considered as an additional source of market failure
Understanding of organisational failures	• traditionally neglected (orthodox) • now understood as a consequence of asymmetric information within internal principal-agent relationships (agency)	• treated in same manner as market failures • understood as a consequence of bounded rationality and opportunism within internal contractual relationships • cognitive biases may be considered as an additional source of organisational failure
Ideas relevant to the barriers debate	• information as a public good • asymmetric information • search, experience and credence goods • pre-contractual opportunism and adverse selection • post-contractual opportunism and moral hazard • appropriability and split incentives	• bounded rationality, satisficing, routines • market transaction costs, including search, bargaining, decision, supervision and enforcement costs • organisational transaction costs, including monitoring, performance measurement, information management and logistics costs • cognitive biases, certainty effect, loss aversion and inertia

Figure 2.1 illustrates how the preceding sections have extended the orthodox model by modifying key assumptions, while Table 2.4 summarises and compares the core ideas that have been introduced. The remainder of this chapter will demonstrate how each of these ideas may improve understanding of the barriers to energy efficiency. Throughout what follows, a primary objective will be to relate the abstract concepts listed in Table 2.4 to the more concrete observations of energy efficiency practitioners.

5. A TAXONOMY OF BARRIERS TO ENERGY EFFICIENCY

Studies of barriers to energy efficiency vary greatly in the extent to which they employ the formal concepts described above. Some studies discuss barriers with minimal reference to formal theory (Hirst and Brown, 1990); some draw upon orthodox ideas but reject or overlook transaction cost and behavioural approaches (Jaffe and Stavins, 1994); and some make reference to concepts such as transaction costs, but do not develop their full implications (Sanstad and Howarth, 1994a). The situation is further complicated by the blurring of boundaries between different approaches. Not only is there a great deal of overlap between the orthodox, agency, transaction cost and behavioural perspectives, but the concepts used in the barriers debate frequently subsume more than one of these ideas and refer to phenomena that can be explained by more than one factor.

The following sections describe and explore the six barriers to energy efficiency identified earlier in Table 1.1. In each case, the proposed barriers represent a potential answer to one or more of the following questions:

- Why do organisations impose very stringent investment criteria for projects to improve energy efficiency?
- Why do organisations neglect projects that appear to meet these criteria?
- Why do organisations neglect energy efficient and apparently cost-effective alternatives when making broader investment, operational, maintenance and purchasing decisions?

Each of the proposed barriers may be understood with reference to one or more of the concepts discussed in the preceding sections. For example, 'hidden' costs may be understood with reference to both the economics of information and TCE. Hence, while the proposed classification of barriers is based upon categories that are widely used and understood within the energy efficiency literature, the description and evaluation of these barriers draws upon more formal ideas. Furthermore, it is argued that several of these ideas are relevant to each of the proposed barriers. Table 2.5 repeats the list of six barriers and summarises their implications. The subsequent sections discuss the nature, operation and consequences of each barrier in turn, drawing upon one or more of the concepts summarised in Table 2.4. Each section also indicates how such barriers may be identified within case study research.

Table 2.5 A taxonomy of barriers to energy efficiency

Barrier	**Claim**
Risk	The short paybacks required for energy efficiency investments may represent a rational response to risk. This could be because such investments represent a higher technical or financial risk than other types of investment, or that business and market uncertainty encourages short time horizons.
Imperfect information	Lack of information on energy efficiency opportunities may lead to cost effective opportunities being missed. In some cases, imperfect information may lead to inefficient products driving efficient products out of the market.
Hidden costs	Engineering-economic analyses may fail to account for either the reduction in utility associated with energy efficient technologies, or the additional costs associated with them. As a consequence, the studies may overestimate energy efficiency potential. Examples of hidden costs include overhead costs for management, disruptions to production, staff replacement and training, and the costs associated with gathering, analysing and applying information.
Access to capital	If an organisation has insufficient capital through internal funds, and has difficulty raising additional funds through borrowing or share issues, energy efficient investments may be prevented from going ahead. Investment could also be inhibited by internal capital budgeting procedures, investment appraisal rules and the short-term incentives of energy management staff.
Split incentives	Energy efficiency opportunities are likely to be foregone if actors cannot appropriate the benefits of the investment. For example, if individual departments within an organisation are not accountable for their energy use they will have no incentive to improve energy efficiency.
Bounded rationality	Owing to constraints on time, attention, and the ability to process information, individuals do not make decisions in the manner assumed in economic models. As a consequence, they may neglect energy efficiency opportunities, even when given good information and appropriate incentives.

5.1 Risk

Both high discount rates for energy efficiency investments and the rejection of particular energy efficient technologies may represent a rational response to risk. For example, if there is doubt that a business will survive over the next three years, stringent investment criteria may be entirely appropriate. Risk may derive from a range of sources, including overall economic trends (for example, inflation, interest rates), potential changes in government policy, trends in input and output markets (for example, fuel and electricity prices), financing risk (for example, the anticipated reaction of capital markets to increases in borrowing) and the technical risks associated with individual technologies (for example, unreliability). These risks may be expected to vary with the individual country, sector, business and technology and to change over time. Furthermore, perceptions of risk by the relevant decision-makers may depart from those suggested by economic models (Tversky and Kahneman, 1991).

While risk is multidimensional, what matters for the barriers debate is the potential impact of real or perceived risks on energy efficiency investment, as opposed to other forms of investment. In other words, are there any reasons why energy efficiency investments or energy efficient alternatives should carry a higher risk than other forms of investment, and therefore be systematically overlooked?

The first possibility is that energy efficient technologies are subject to greater technical risks. For example, if technology is (or is perceived to be) unreliable, the risk of breakdowns and disruptions may outweigh any potential benefits from reduced energy costs. Such risks are particularly associated with new and unfamiliar technologies and these are commonly the subject of government-funded demonstration programmes which aim to increase confidence and disseminate information and awareness among potential adopters. However, many of the technologies that are included in engineering-economic models and recommended in energy efficiency publications are well proven, reliable and widely used. These include, for example, energy efficient lighting, condensing boilers, thermal insulation, energy efficient motors, thermostatic radiator valves and lighting controls. In most applications the technical risk associated with these technologies is very small, which makes it unlikely that risk can provide a rational reason for their rejection in the majority of cases. In the empirical research reported in chapters 3 to 7, an attempt is made to minimise the importance of technical risk by focusing on technologies that are well established, proven and widely recommended in the best-practice literature.

A second possibility, suggested by Sutherland (1991), is that energy efficiency investments require higher hurdle rates because they are 'illiquid' and irreversible, with limited scope for diversifying risks. The comparison here is with investments in financial instruments, such as stocks and bonds, which are highly liquid since they can be easily bought and sold (Golove and Eto, 1996, p. 44; Johnson, 1994). In comparison, energy efficiency investments are normally embedded within buildings and equipment, costly to remove and with limited scope for subsequent resale. Since they must generally be retained, regardless of their performance, they carry a greater risk than investment in other types of assets, which suggests that the value of future benefits should be discounted more highly. A similar argument is provided by TCE, in that highly specific assets with limited scope for redeployment are generally associated with higher risks (Golove and Eto, 1996, p. 44). But while this argument may account for the differing treatment of 'liquid' and 'illiquid' assets, it fails to account for the differing treatment of comparable assets. For example, why should cost saving energy efficiency investments be subject to more stringent investment criteria than investment in new production plant, when the latter is equally illiquid and irreversible? The scope of this argument may therefore be limited.

Hassett and Metcalf (1993) and Johnson (1994) present a similar argument to Sutherland, which is based on the concept of 'real options theory' (McDonald and Siegel, 1986; Dixit and Pindyck, 1994; Ostertag, 2003). They argue that the combination of uncertainty in the future course of energy prices, capital costs and technical change, coupled with the irreversibility of energy efficiency investments leads to an optimal rate of return which is higher than conventional investment models predict. One reason for this is that there is an opportunity cost in acting today, rather than delaying the decision and resolving some uncertainties. The possibility of delaying the decision represents an option, which has a value in proportion to the degree of uncertainty on investment returns. But since this option is no longer available once the investment has been made, the full cost of the investment should include the cost of foreclosing the option. This leads to a testable implication that required rates of return should be positively correlated with the degree of uncertainty of future returns (Metcalf, 1994).

Sanstad et al. (1995) have criticised the Hasset and Metcalf model on three grounds. First, it fails by some distance to account for the observed discount rates for energy efficiency investments. They demonstrate that the 'option value multiplier' in Hasset and Metcalf's model falls off rapidly as the assumed discount rate increases, and has only a limited effect on the required rate of return (for example, increasing from 15 per cent to 17.4 per cent). Second, the model fails to account for the potential *costs* of delaying energy efficiency investments (Howarth and Sanstad, 1995, p. 105). For example, it is much more costly to retrofit heat recovery systems than to include them when a plant or building is designed. Since most decisions relevant to energy efficiency involve a choice between efficient and inefficient options within an investment that is being made for other purposes, the scope of the model is seriously circumscribed. Third, the model has the usual limitations of orthodox theory in that it assumes investors are fully informed about relevant alternatives and able to solve sophisticated optimisation problems. This compares poorly with models that take account of limitations on decision-making, and is inconsistent with empirical studies of energy demand behaviour.

Given these considerations, the argument that high discount rates can be considered a rational response to risk for *all* types of energy efficiency investment does not seem plausible. The quantitative predictions of the models fail on their own terms, quite apart from the implausibility of the behavioural assumptions and the limited range of investment decisions for which they seem applicable. However, business, regulatory or technical risk may be a relevant and important factor in individual cases and the importance of this should be investigated within the case study research.

5.2 Imperfect Information

The importance and policy implications of imperfect information is one of the central issues in the barriers debate. The primary claim is that, for a variety of reasons, individuals lack adequate information on either individual energy efficiency opportunities or on the energy performance of different technologies. This leads them to make sub-optimal decisions based on provisional and uncertain information, and consequently to under-invest in energy efficiency. Since imperfect and asymmetric information are central to the orthodox understanding of market failures, the existence of imperfect information is claimed to justify policy interventions to improve information such as energy labelling. Huntington et al. (1994, p. 796) argue that '…information problems taking different forms are the principal source of market failures that account for the "gap" in energy efficiency investments'.

Imperfect information in energy service markets

The information relevant to energy efficiency decisions may usefully be grouped into three categories:

- information on the level and pattern of current energy consumption and the comparison of this with relevant benchmarks;
- information on specific energy-saving opportunities, such as the retrofit of thermal insulation; and

- information on the energy consumption of new and refurbished buildings, process plant and purchased equipment, allowing choice between efficient and inefficient options.

The availability of information on current energy consumption will depend upon the information content of utility bills, the level of sub-metering, the availability of relevant benchmarks, the use of computerised information systems, the time devoted to analysing consumption information and so on. Most of these will be associated with investment, operational and staff costs which may best be understood as a particular category of transaction cost (Table 2.3). These costs are of fundamental importance to the barriers debate and are discussed more fully in the next section.

The availability of information on energy-specific investment opportunities will depend upon two factors: the extent to which organisations have evaluated energy efficiency opportunities through measures such as energy audits; and the availability of information on the costs and performance of specific energy-saving technologies. The first of these involves costs for the organisation (whether the audits are conducted in-house or by external consultants), and these may also be understood as a form of organisational transaction cost. The information provided by energy audits is private and customised, so should be relatively free of public good problems, but as Goldstone (1995) points out:

> '... on the one hand, since the benefits of customised information are internal to the customer, they should be fully provided by markets. On the other hand, the customer can't determine the benefits until after he receives the information, at which point he already has it. It is thus difficult to see, even assuming rationality, how the market would yield this result. So in general we expect that the market will under-supply or overprice relevant information...'.

Information on specific energy efficient technologies should be available in the marketplace, but the cost, quality and accuracy of this information may vary widely between different technologies. In some cases, this information may be subject to public good problems, and for new and unfamiliar energy efficiency technologies there may be a case for publicly-funded information programmes and demonstration schemes.

The extent to which information is available on the energy consumption of new and refurbished buildings and purchased equipment will depend very much upon the characteristics of the relevant market. Energy efficiency is not a stand-alone product in these markets, but a secondary and often relatively unimportant feature of a wide range of goods and services (Golove and Eto,

1996, p. 29). Energy services are delivered through a combination of energy commodities, building and transport infrastructures and energy conversion technologies, and the decisions made on the specification, design, purchase, installation, operation, repair and maintenance of a wide range of technologies will influence the overall energy efficiency performance (Golove and Eto, 1996, p. 29). The informational problems associated with energy service markets will therefore be specific to individual technologies and services (for example, motors, lights, buildings, pumps, appliances). If information barriers are significant, the market signals in favour of more energy efficient products and services are likely to be relatively weak.

Asymmetric information in energy service markets

Energy service markets are likely to be characterised by asymmetric information between producer and purchaser and between market intermediaries at different stages along the supply chain. As indicated earlier, the importance of this will depend upon the variance in product quality (particularly in relation to energy efficiency), the frequency of purchase relative to changes in underlying characteristics and the search costs entailed in obtaining relevant information. It is useful here to compare the relative importance of asymmetric information for the purchase of energy commodities such as gas and electricity, compared to the purchase of energy efficiency products or energy efficiency equipment which may allow the same level of energy service to be obtained at lower levels of energy consumption (Hewett, 1998, p. 2.8). For example, the same level of thermal comfort may be obtained by using more fuel within an existing heating system, or by investing in loft insulation, cavity wall insulation, draught stripping and/or double glazed windows.

Energy commodities represent a simple, unchanging, easy to understand and homogeneous product which is purchased from a small number of large, well established and trusted firms. Purchases are made regularly, market information is widely available and 'performance' is judged largely on price. Hence, energy commodities can be considered a search good with relatively low search costs.[17] In contrast, delivering the same service through energy efficiency investment requires the purchase of one or more complex, heterogeneous and unfamiliar goods from markets with multiple suppliers and intermediaries. Since the lifetime of such products is long (for example, 15 years), the purchases are infrequent and the rate of technical change is rapid relative to the purchase interval (Hewett, 1998, p. 2.11). For example, technologies such as condensing boilers, building energy management systems and electronic ballasts have improved enormously over the last ten years.

In contrast to energy commodities, energy efficiency may only be considered a search good when the energy consumption of a product is clearly and unambiguously labelled and when the performance in use is insensitive to installation, operation and maintenance conditions. But for many goods, the information on energy consumption may be missing, ambiguous or hidden, and the search costs will be relatively high. In the absence of standardised performance measures or rating schemes, it may be difficult to compare the performance of competing products. Even when rating schemes are available, the performance in use may depart significantly from the rated performance – for example when technologies are operated on part load or are inadequately maintained. Also, customers may have great difficulty in evaluating the performance claims of technology suppliers, or may be suspicious of these claims. In these circumstances, energy efficiency is better described as an experience good.

In practice the performance of technologies such as control systems, motors and variable speed drives may be very difficult to evaluate even after purchase. In most cases, the evaluation of energy performance would require low-level electronic sub-metering, adjustment for variable factors such as occupancy and weather, and careful analysis of consumption patterns over time. If this is not done, the purchaser will lack feedback on the consequences of different purchase decisions, with the result that energy use will be relatively invisible (Hewett, 1998, 2.11). For example, Kempton and Montgomery (1982) have compared the information value of the average household energy bill to that of receiving a single monthly bill from the supermarket for 'food'. Taken together, these features tend to make energy efficiency closer to a credence good than an experience good and hence more subject to market failure. Thus, to the extent that energy supply and energy efficiency represent different means of delivering the same level of energy service, the latter is likely to be disadvantaged relative to the former. The result is likely to be over-consumption of energy and under-consumption of energy efficiency (Hewett, 1998, p. 2.8).

The choice between energy supply and energy efficiency is most relevant to energy-specific investment opportunities, such as thermal insulation. But in practice, a more common choice is between an energy efficient or inefficient product or service, when a decision is required anyway – for example, replacing an existing boiler which has come to the end of its life. What is relevant here is the availability of information on the energy performance of the product (and the consequent savings in operating costs) as compared to the availability of information on other attributes of the product, such as capital costs. This in turn will depend upon the relative importance of energy consumption as compared to the other services delivered by the product. For example, thermal efficiency may be expected to be an important and visible attribute of a boiler. But energy efficiency may be expected to be very much a secondary attribute of other products such as buildings, and is likely to be determined by a wide range of design and operational factors, the net effect of which may be difficult to assess. In the absence of clearly specified and comparable performance information, energy efficiency considerations are likely to be easily outweighed by other more visible features. Hence, even if energy efficiency is valued by the consumer, the lack or cost of information may prevent this preference from being exercised: '...Faced with good information on capital costs and poor information on operating costs, consumers may rationally and systematically choose the low capital option' (Eyre, 1997, p. 31).

Adverse selection in energy service markets

In some circumstances, asymmetric information in energy service markets may lead to the adverse selection of energy inefficient goods. Take housing as an example (Jaffe and Stavins, 1994, p. 809). In a perfect market, the resale value of a house would reflect the discounted value of energy efficiency investments. But asymmetric information at the point of sale tends to prevent this. Buyers have difficulty in recognising the potential energy savings and rarely account for this when making a price offer. Estate agents have greater resources than buyers, but similarly neglect energy efficiency when valuing a house. Since the operating costs of a house affects the ability of a borrower to repay the mortgage, they should be reflected in mortgage qualifications. Again, they are not. In all cases, one party (for example, the builder or the seller) may have the relevant information, but transaction costs impede the transfer of that information to the potential purchaser. The result may be to discourage house builders from constructing energy efficient houses, or to discourage homeowners from making energy efficiency improvements since they will not be able to capture the additional costs in the sale price.

The same processes are at work in a range of energy services markets. In some cases, producers may be unable to market desirable technologies since consumers are unable to observe their characteristics prior to sale (Howarth and Sanstad, 1995, p. 106). In other cases, information asymmetries may create incentives for producers or suppliers to act opportunistically. For example, the energy efficiency of commercial buildings depends heavily on the detailed features of heating, ventilation and controls such as Building Energy Management Systems (BEMS). But in comparison to highly visible features such as outward form and aesthetics, the performance of building services equipment is extremely difficult for the customer to observe. Substitution of an inefficient or oversized piece of equipment in place of efficient equipment could be relatively easy. In an analogous manner to

Akerlof's second-hand car market, the result may be inefficient products driving efficient products off the market.

Overcoming information problems in energy service markets

Problems of imperfect and asymmetric information may provide a rationale for policy intervention, but the appropriate form of intervention may be contested. Sutherland argues that labelling schemes are legitimate as they directly address the information market failure, but minimum energy efficiency standards are not legitimate since they prevent some customers from exercising their preferences for low capital cost/high energy cost products (Sutherland, 1996, p. 364). But since minimum standards are commonly applied to other types of credence goods such as pharmaceuticals, there may be a case for extending their use to certain types of energy efficient products. To justify such a policy, it would be necessary to take into account the costs to producers of shifting towards more efficient goods, the benefits to consumers of lower energy costs, the extent to which high discount rates represent true time preferences as opposed to other forms of market failure, and the likely effectiveness of different types of policy.

From a behavioural and transaction cost perspective, what matters is not merely the availability of information but the *credibility* of the source (Stern, 1984, p. 43). A classic study by Craig and McCann (1978) illustrates this point. One thousand New York households were sent a pamphlet describing how to save energy. Half the households received the mailing from the local electric utility, and the other half from the state regulatory agency. The following month, households that had received pamphlets from the agency used about 8 per cent less electricity than those that had received the identical pamphlets from the utility.

Perceptions of credibility will depend upon the expertise and perceived trustworthiness of the source (for example, private, governmental, charity), past experience with the source, the nature of interactions with the source, recommendations from colleagues, and recommendations from a wide range of contacts within professional and social networks. Practical experience with energy efficiency programmes suggests that interpersonal contacts and recommendations count for significantly more than labels, pamphlets and paper qualifications (Stern, 1994, p. 67). This is because information from close associates stands out from the mass of available information and attracts attention. Most interpersonal contacts are made through existing professional and social networks which play a fundamental role in transmitting information and establishing trust. For example, US household energy efficiency programmes achieved greater success when implemented through community groups which had established credibility through extensive personal contacts (Stern, 1986). In the organisational context, networks organised through trade associations are likely to be similarly effective.

A second issue is the form in which information is presented since this will affect the transaction costs of acquiring, understanding and applying the information. Poorly presented information requires more cognitive effort to assimilate and hence is more easily ignored. For example, US evaluations of energy efficiency programmes showed that people frequently ignored information even when it was freely available (Ester and Winett, 1982). Hence, while the search costs were zero, the transaction costs were still prohibitive. To be effective, information needs to be clear and simple, available close in time to the relevant decision and vivid.[18] Also, specific and personalised information such as that provided by energy audits, should be more effective than general information on cost saving opportunities, while a demonstration of tangible success with a technology is likely to have more persuasive power than a sales pitch. A final point is that feedback should be available on the beneficial consequences of previous energy decisions if subsequent energy efficiency measures are to be encouraged (Seligman et al., 1981).

Summary

In summary, problems of imperfect information are likely to pervade energy service markets and could potentially explain a substantial proportion of the efficiency gap. First, the acquisition of information through measures such as metering and audits involves investment and transaction costs which may not be taken into account in engineering-economic models. Second, the search costs for energy efficient products are likely to be much greater than those for energy commodities, creating a systematic bias against energy efficiency. Third, energy efficiency has the characteristics of a credence good, which makes it particularly vulnerable to information market failure. And fourth, asymmetric information in energy service markets may sometimes lead to the adverse selection of energy inefficient goods.

The appropriate policy response to these market failures is contested. While information programmes appear to be the most obvious approach, minimum energy efficiency standards may be more effective in some instances. If information programmes are to be employed, both the manner in which information is presented and the credibility of the source are important.

The importance of information costs suggests that there is a considerable overlap between the 'imperfect information' barrier and the 'hidden cost' barrier. The next section clarifies these overlaps in more detail.

5.3 Hidden Costs

Hidden costs represent the most important and influential explanation for the 'efficiency gap'. The claim is that engineering-economic studies fail to account for either the reduction in utility associated with energy efficient technologies, or the additional costs associated with their use (Nichols, 1994). As a consequence, the studies tend to overestimate energy efficiency potential.

Components of hidden costs

In the energy economics literature, the term 'hidden costs' refers to any costs which are not conventionally included within engineering-economic models. Three possible sources of hidden costs are:

- the general *overhead* costs of energy management;
- the costs which are *specific* to an individual investment in energy efficiency, or the choice of an energy efficient option; and
- the potential loss of *utility* associated with energy efficient choices.

Table 2.6 identifies the possible components of these costs in more detail.

The sources of costs listed in Table 2.6 include some which would be regarded as search costs within the economics of information, some (a broader group) which would be regarded as transaction costs within TCE, and some which would be regarded as neither. Each of the cost categories may be both real and significant and each may partly account for the gap between real-world investment behaviour and the predictions of engineering-economic models. But the different categories vary widely in their likely importance in particular instances, the extent to which they are specific to individual sites and technologies, the ease with which they can be quantified and incorporated into energy models, the extent to which they can be reduced by changes in organisational procedures and routines, and their relevance for public policy. As a result, the overarching category of 'hidden costs' needs to be disaggregated into its individual components so that the determinants, relevance and relative importance of each component can be assessed. It is tautologous to assert that hidden costs *must* be present if organisations are not adopting particular energy efficient technologies. Instead, it is necessary to demonstrate what (if any) those costs are, why they are important, what is determining them and whether and how they could be reduced.

Some hidden costs could be considered part of the *production* cost of energy efficiency and could in principle be included in engineering-economic models (Ostertag, 1999). Examples include design fees for large items of plant, the civil engineering costs associated with installing a CHP unit, the costs of re-routeing pipework, the reinforcement costs associated with connecting a CHP unit to the grid, the costs of rebuilding a flue after a condensing boiler has been installed, the cost of new light fittings to accommodate compact fluorescents, and the cost of production interruptions during equipment installation. These costs are site-specific and difficult to estimate, so they may be easily overlooked. But they represent real costs and organisations can be expected to take them into account when appraising investment opportunities.

Table 2.6 Empirical perspective on the components of hidden costs

Sub-category	Examples
General overhead costs of energy management	• costs of employing specialist people (e.g. energy manager) • costs of energy information systems (including: gathering of energy consumption data; maintaining sub-metering systems; analysing data and correcting for influencing factors; identifying faults; etc.) • cost of energy auditing
Costs involved in individual technology decisions	• cost of: i) identifying opportunities; ii) detailed investigation and design; iii) formal investment appraisal • cost of formal procedures for seeking approval of capital expenditures • cost of specification and tendering for capital works to manufacturers and contractors • additional staff costs for maintenance • costs for replacement, early retirement, or retraining of staff • cost of disruptions and inconvenience
Loss of utility associated with energy efficient choices	• problems with safety, noise, working conditions, service quality etc. (for example, lighting levels) • extra maintenance, lower reliability

A second group of costs relates to the inferior performance of energy efficient technologies along a number of dimensions other than energy consumption. For example: an energy efficient production process may lead to increased noise; the installation of cavity wall insulation in an old building may encourage damp; a variable speed drive may require extra maintenance or require new skills and tools; an energy efficient motor may be less reliable; the lighting quality from compact fluorescents may be less desirable than that from incandescent bulbs; and so on (Golove and Eto, 1996, p. 16). While these considerations clearly apply to energy-specific investment opportunities, they are likely to be even more important for investments where energy efficiency is only one of a number of attributes under consideration. Again, in principle these costs could be incorporated within engineering-economic models, but this may be difficult to achieve in practice.

A third group of costs corresponds to the search costs identified in the economics of information and discussed in the previous section. These include the cost of identifying suppliers and obtaining information on price, quality and terms of trade. As argued above, these costs are strongly influenced by the characteristics of particular energy service markets and by the nature of energy efficiency as a good. They are determined in part by factors outside the control of the adopting organisation, such as the existence or otherwise of standardised labelling schemes, and in part by organisational procedures for information gathering, specification, purchasing and procurement. Search costs are therefore influenced by a mix of factors both internal and external to the organisation, and public policy may have greater scope for influencing the latter than the former.

Search costs represent a subset of the broader category of transaction costs. These include all the organisational costs associated with establishing and maintaining an energy management scheme, investing in specific energy saving technologies, and implementing specific energy efficient options within broader investment programmes (for example, choosing energy efficient motors in preference to standard motors). In contrast to the production costs and loss of utility discussed above, transaction costs depend closely upon organisational and contractual structures, procedures, incentives and routines. This makes them much more difficult to incorporate within models which represent costs purely in relation to individual technologies (Ostertag, 1999).

To summarise the above, Table 2.7 provides a theoretical perspective on the components of hidden costs which complements the empirical perspective provided by Table 2.6. This divides hidden costs into four categories: production costs; loss of utility; market transaction costs; and organisational transaction costs. The first two of these categories may be considered as real and unavoidable costs which have no implications for public policy, while the last two categories are contingent upon the relevant market, contractual and organisational structures and hence may in some circumstances be lowered through public or private actions. Identifying the size and determinants of the these transaction costs is likely to be difficult, but box 2.1

highlights four important features of these costs as they relate to energy efficiency.

Table 2.7 Theoretical perspective on the components of hidden costs

Sub-category	Influenced by	Examples
Hidden production costs	• Attributes of technology • Site-specific factors	Civil engineering costs, grid reinforcement costs, production interruptions
Loss of utility	• Attributes of technology • Site-specific factors	Increased noise, reduced service quality
Market transaction costs	• Features of primary and secondary markets for information. • Organisational procedures for external transactions	Search costs for gathering and assimilating information regarding product quality; cost of specification and tendering; bargaining and negotiation costs; legal advice; etc.
Organisational transaction costs	• Organisational procedures for internal transactions	Monitoring and control costs; decision-making costs; costs of establishing, maintaining and running energy information systems; etc.

Box 2.1 Transaction costs and energy efficiency

- Transaction costs need not increase in proportion to the volume of the transaction. So for example, the transaction costs associated with identifying and purchasing an energy efficient motor will form a declining proportion of total life-cycle costs as the size of the motor increases.
- Transaction costs are likely to be incurred for both energy efficient and inefficient choices. So if a replacement motor is required, there will be transaction costs for purchasing both energy efficient and

inefficient models, and what matters is the difference between the two.

- Transaction costs may only accrue once when an organisational routine is changed, such as when a shift is made from purchasing standard to high frequency fluorescent lighting. So comparing an exceptional situation with an established organisational routine may overstate the costs involved.
- Transaction costs may decrease over time as a result of learning effects as knowledge becomes embedded within individual and organisational routines. So they should not be treated as fixed and unchanging.

Source: Ostertag (1999).

Identifying hidden costs

There are two approaches to assessing the importance of hidden costs. A *cardinal* approach would attempt to quantify the hidden costs associated with a specific energy efficient investment, while an *ordinal* approach would attempt to identify instances where the hidden costs associated with an energy efficient technology are comparable with those of an inefficient technology, but where the latter are still predominantly chosen. Both are challenging, and research designs would need to be comprehensive to ensure that all relevant factors were taken into account.

Cardinal approaches are rare in the literature. One example is the 1994 study by Hein and Blok, which estimated the staff costs associated with collecting information, making decisions and monitoring the performance of different types of energy efficiency investments within large, energy intensive firms. The transaction costs were dominated by search costs and typically formed between 3 per cent and 8 per cent of the total investment cost. This suggests that transaction costs were relatively small for this type of organisation, but they could be much more important for small-scale investments or for smaller, less energy intensive firms. Transaction costs are therefore relative to the technical and organisational context.

An example of a ordinal approach is provided by Levine et al. (1994), who studied a number of energy efficient technologies which were commercially available, identical to inefficient technologies in the quality of service provided, highly cost-effective, and apparently free of any hidden costs. Despite this, inefficient technologies with a somewhat lower capital cost were generally preferred. The technologies studied included ballasts in commercial fluorescent lights, residential refrigerators and standby power in colour televisions. In these cases, hidden costs appear unable to account for the observed efficiency gap.

For organisations, one of the most important sources of hidden costs is likely to be the overhead costs of energy management. These have rarely been subject to serious academic study, but are frequently cited by industry as the biggest obstacle to cost-effective investment. For example, negotiated agreements between government and industry in the UK require the implementation of energy efficiency projects with paybacks as short as three years (Sorrell, 2002) and the primary reason given by UK industry for the use of these strict investment criteria is the management time required to identify and implement such projects (ETSU, 2001). These claims should be treated with suspicion, given the information asymmetry between government and industry and the incentive for industry to exaggerate the importance of hidden costs in order to reduce the stringency of the negotiated agreements. But they demonstrate that management time is a predominant concern and that it is accepted by the UK government as an adequate explanation of the efficiency gap.

Some simple calculations can demonstrate the potential importance of overhead costs for a typical organisation. Best-practice literature in the UK recommends that a sum equivalent to 5 per cent of an organisation's annual energy expenditure be reserved for dedicated energy efficiency investment (Energy Efficiency Office, 1993). For a site with a £1m bill, this would equal £50k. But the annual salary costs for a full-time energy manager may be £30k to £40k, or 60 to 80 per cent of the annual investment budget. In this context, stringent payback criteria for such investment projects may be justified as a means to recover the salary overheads – for example, by transforming a five-year payback into a three-year payback. Furthermore, organisations with small energy bills would only be able to devote a fraction of staff time to energy efficiency and would be unlikely to develop the relevant energy management skills. However, much of energy management could be seen as an essential function necessary for the functioning of the institution and the comfort of employees – for example, overseeing maintenance, running building energy management systems and negotiating with energy suppliers (Box 2.2). Other tasks, such as internal reporting, maintaining and upgrading information systems, training, marketing and awareness-raising are unlikely to produce direct savings, but are essential to creating a climate that is supportive of attempts to improve energy efficiency (Energy Efficiency Office, 1993, p. 38). As a result, it appears unreasonable to require *all* the salary overheads to be recovered solely from the returns on individual energy efficiency projects. At the same time, it appears equally unreasonable for *none* of the salary overheads to be recovered from such projects. Best practice guidelines suggest a figure in the range 10–40% (Energy Efficiency

Office, 1993, p. 39), but the optimal proportion is likely to vary between organisations.

Summary

In sum, the majority of energy efficiency investments are likely to be associated with some form of hidden cost, and these costs could potentially explain a portion of the efficiency gap. A proportion of these costs may be considered part of the production cost of energy efficiency and hence provide no rationale for policy intervention. But a proportion may be understood as market or organisational transaction costs and hence could potentially be reduced through organisational changes or public policy. Hence, while the assertion that the hidden costs can explain all the efficiency gap is merely tautologous, the assertion that hidden costs are unimportant is equally likely to be wrong. The truth should lie somewhere between the two, but the relative importance of different categories of cost is likely to vary between individual technologies and organisations. It appears likely, however, that the salary overheads associated with energy management will be a major obstacle for many organisations, and especially for those with relatively small energy bills.

Box 2.2 Tasks and responsibilities of an energy manager

- formulating and implementing an energy policy
- purchasing energy at competitive rates
- introducing, maintaining and operating energy information systems
- reporting energy information appropriately to accountable staff and senior managers
- raising and maintaining energy awareness among staff
- introducing and maintaining effective 'housekeeping' and operating practices
- identifying and fulfilling training needs for energy-related skills
- identify cost-effective opportunities for improving energy efficiency in both existing and new plant and buildings
- formulating an investment programme for energy efficiency
- introducing and maintaining review procedures for establishing the value for money of energy management activities

Source: Energy Efficiency Office, 1993, p. 18.

5.4 Access to Capital

A commonly cited barrier within the energy efficiency literature is the lack of access to capital (Hirst and Brown, 1990; Eyre, 1997). This is particularly applicable to the domestic sector, where low-income households have limited access to credit and can only borrow at high interest rates. This may prevent energy efficiency projects with a high rate of return from being undertaken. The orthodox response to this argument is that while inability to access capital may constitute a barrier, it need not imply a failure in capital markets. Capital should be allocated to projects with the highest risk-adjusted rate of return, and groups such as low-income households tend to be high-risk borrowers (Sutherland, 1996). An alternative view is that the transaction costs entailed in investigating the creditworthiness of individual households are sufficiently high to diminish the economic viability of such loans (Golove and Eto, 1996). As a result, policy interventions to overcome this barrier in the household sector tend to be justified primarily on equity grounds.

The situation is more complex for private sector firms. Here, the 'access to capital' problem has two components:

- insufficient capital through internal funds, and potential difficulty in raising additional funds through borrowing or share issues; and
- neglect of energy efficiency within internal capital budgeting procedures, combined with other organisational rules such as strict requirements on payback periods.

Both of these are the subject of a voluminous theoretical and empirical literature (Stein, 2001; Schleifer and Vishny, 1997; Myers, 2001). This work relies primarily upon agency theory and develops and tests a wide range of frequently contradictory hypotheses. Some key points relevant to both external and internal capital markets are discussed below.

Accessing external sources of capital

Within private sector firms, restrictions on capital are often self-imposed. Here firms appear reluctant to borrow money to finance low-risk energy efficiency projects with rates of return that significantly exceed their weighted average cost of capital (WACC).[19] In some cases, this reluctance appears to result from the perceived risk of increasing the ratio of loan finance to equity finance – termed *gearing* (Ross, 1986; Mclaney, 1994). Loan finance should be valuable up to a point, since it tends to be cheaper than equity – historically, the expected returns from equities are higher than those from loan stocks, and loans tend to have a more favourable tax treatment. But loan finance carries risks in that it imposes obligations to meet annual interest charges and to repay the principal. In contrast to share dividends, these are fixed obligations and are not at the firm's discretion. High levels of gearing may therefore expose the firm to the risk that it will not be able to meet its payment obligations should it experience a downturn in business.

With loan finance, the lenders have the legal right to enforce payment of the interest and repayment of the capital, using the assets of the company as security. In contrast, ordinary shareholders do not have the right to enforce the payment of a dividend. This situation means that high levels of gearing may expose the shareholders to greater risk as all the firm's profits could be eaten up in repayments to lenders. As result, shareholders may demand higher returns as compensation. Furthermore, high levels of gearing may also expose the *lenders* to greater risk, since the asset value may be insufficient to pay off the outstanding loans should the firm go out of business. Hence, lenders may also demand higher interest payments on loans as the level of gearing increases. The level of risk will depend in part upon the resale value of the assets (in TCE terminology, the degree of asset specificity) and also upon the proportion of total costs within the firm which are fixed – termed *operating gearing*. The risk of bankruptcy is particularly high for firms which are highly geared in both operating and capital terms, since relatively small fluctuations in the level of sales can have a dramatic effect on profits (Mclaney, 1994, p. 273). The net result is that, while loan finance may reduce a firm's cost of capital at low levels of gearing, it may increase risk and raise a firm's cost of capital at high levels of gearing. Management may therefore restrict the level of gearing to a level they feel comfortable with.

This traditional view of an 'optimal' level of gearing was challenged by Modigliani and Miller (1958), who showed that, given a set of assumptions about the operation of the capital market, the advantages of cheaper loan finance should be exactly offset by the increasing cost of equity. As a result, the WACC should be independent of the level of gearing and should depend solely upon business risk and future cash flows. But this model effectively assumes that the transaction costs within capital markets are zero (Furubotn and Richter, 1997). Also, this theoretical result is not supported by the empirical evidence, which shows a reluctance to increase gearing beyond a particular level (Myers, 2001, p. 275).

Other perspectives on the use of capital markets emphasise information asymmetry and agency problems. For example, Jensen (1986) argues that increased gearing may be in the interest of shareholders if it lowers the cost of capital, but may not be desirable from the perspective of company directors because it imposes a discipline which they may prefer to avoid. This could dissuade directors from using external funds to finance cost-effective investments. Similarly, Myers and Majluf (1984) emphasise how a reliance upon external finance may be interpreted by investors as a signal that the existing assets are overvalued. It is commonly observed that an attempt to raise additional equity finance or to increase the level of gearing can weaken a firm's financial rating and drive down share and bond prices (Asquith and Mullins, 1986). Since debt imposes both greater risk on the firm and greater discipline upon managers, it should have a smaller impact than share issues. But in all cases, the cost of obtaining additional capital may exceed the average cost of the existing debt/equity mix (Ross, 1986). The prediction, therefore, is that firms will: a) prefer internal to external finance; b) prefer debt finance to equity; and c) avoid high levels of gearing (Myers, 2001).

Since most energy-specific investment opportunities involve relatively small amounts of capital, they should have very little impact on the level of gearing for the firm as a whole (particularly for larger companies). But since borrowing requirements and 'financing risk' are likely to be assessed for the firm as a whole, and not for individual investments, the effect may be to restrict the overall capital budget for investment, including that for energy efficiency. This effect may be exacerbated by internal capital budgeting procedures which give a lower priority to discretionary cost saving investments (see below), with the result that the dedicated energy efficiency budget is reduced. There are analogous problems with energy efficient options within broader capital investments, such as new buildings. These are typically subject to tight constraints on the capital budget combined with a strong incentive to keep the project within budget. Given the transaction costs associated with seeking (small) additional increments of funding, the budget for energy efficient options (with higher capital costs but lower operating costs) may be squeezed.

In sum, the conventional view that firms should invest in all projects which have a rate of return exceeding the WACC appears over simplistic. There may be very good reasons for not taking on additional debt or raising additional equity, but these require judgements about business risk and the response of the financial market to any increase in gearing. This makes it difficult to assess whether the behaviour of any individual firm is rational, or whether it reflects a failure on the part of management.

Accessing internal sources of capital

Two observations are commonly made in respect of the organisational treatment of dedicated energy efficiency investments. First, such investments tend to be classified as discretionary business maintenance projects, which are given a lower priority than either essential business maintenance projects, such as replacing a failed pump, or strategic business development investments, such as a new manufacturing plant (Department of Energy, 1983). Second, such projects tend to be evaluated using payback rates rather than discounted cash flow analysis, with the required rates of return exceeding those for business development projects (Train, 1985).

A number of hypotheses have been proposed to explain this behaviour, but a common theme is asymmetric information within internal agency relationships, such as those between central and divisional management. The agent (for example, the divisional manager) is closest to a project and is likely to know most about its prospects, but at the same time may have an incentive to misrepresent this information. The principal (for example, central management) cannot observe either the true quality of decision-making or the true profitability of the project. Principals will be unfamiliar with the specific local conditions in which the agent makes her decisions and there will be transaction costs entailed in transmitting the relevant information. This creates the risk that profits will be dissipated into 'managerial slack' – defined as the excess of resources over the minimum required for the task (DeCanio, 1993, p. 909). One method of reducing this slack is to set the 'hurdle rate' for investment projects substantially above the cost of capital to ensure that only highly profitable investments are undertaken (Antle and Eppen, 1985; Narayanan, 1985). Under certain assumptions this may allow the principal to control moral hazard at lower cost. Furthermore, the hurdle rate may be expected to be higher for small investments, since the transaction costs of determining the profitability of such investments are likely to represent a greater portion of the expected savings. Energy efficiency investments frequently fall into this category of small, cost saving investments.

A second explanation could be the strategic priorities of top management. DeCanio (1994) suggests that managers are primarily concerned with ensuring the long-term survival of their organisation, which involves focusing upon dynamic factors such as the introduction of new products and the development of new production facilities. Given severe constraints on time and attention (that is, bounded rationality), the small reductions in costs available from energy efficiency investments could easily be downgraded and overlooked. This is despite the fact that such investments have frequently been shown to have a higher rate of return than large projects which receive more management attention (Ross, 1986). This focus is unlikely to change until energy becomes more of a strategic issue, perhaps through energy prices internalising the external cost of carbon emissions.

The bias towards strict investment criteria may be exacerbated by the incentives on individual managers, including the asymmetry between the risks and rewards of energy efficiency projects. The failure of a project in which an individual had invested considerable effort could be very detrimental to that person's career, while the success of a comparable project could provide a much smaller career boost (DeCanio, 1994, p. 115).

Similarly, managerial advancement may best be achieved through large, strategic projects, while the compensation and prestige for energy management activities may be limited.

Hence, as with accessing external sources of capital, there may be good reasons for imposing strict investment criteria or restricting capital budgets for energy efficiency investments. In both cases, the primary questions for empirical research are: the rationale for such behaviour within the case study organisations; the extent to which this is reproduced in other comparable organisations; and whether such behaviour is a contingent feature of the particular organisational arrangements and hence amenable to change.

5.5 Split Incentives and Appropriability

Split incentives were discussed earlier and were argued to result from asymmetric information and the transaction costs of developing shared savings contracts. While this barrier is most commonly cited in relation to rental housing, it is of much wider applicability.

Landlord–tenant problems may arise in the industrial, public and commercial sectors through the leasing of buildings and office space. For example, in the UK only 10 per cent of commercial property is occupied by the freeholder and 70 per cent is multi-tenanted (Herring et al., 1988). Much of the stock is owned by institutional investors who treat the property purely as an asset, while management is outsourced to property consultants who pay little attention to energy efficiency (Wade et al., 2003). Tenants may have little motivation to improve the performance of an asset they do not own, particularly if they have a short-term lease, while owners will be happy to pass on the energy costs to their tenants. In many cases, tenants will simply pay a fixed pro-rata share of the building's energy bill, which means the savings generated by investment or behavioural changes by one tenant would accrue to all the other tenants as well, thereby diluting the incentive. The problem could be overcome through low-level sub-metering, but this may be costly and it appears to be relatively rare. The landlord–tenant hypothesis suggests an obvious empirical test – whether the energy performance of leased buildings is significantly poorer than other buildings – but there appears to be little research on this topic, and the available evidence is limited (Sutherland, 1996).

Within organisations, the bias towards projects with short-term paybacks may also result from split incentives. It is often the case that managers remain in their posts for relatively short periods of time (DeCanio, 1993, p. 908). In large companies, there may even be a policy of job rotation. But a manager who is in a post for only two or three years has no incentive to initiate investments that have a longer payback period. The incentive structure may therefore be skewed towards projects with rapid returns – although these may prove inferior to others if a full discounted cash flow analysis were performed. As with landlords and tenants, problems of information and transaction costs may prevent the incentive structure from being modified. Statman and Sepe (1985) point to a related issue in that, even without job rotation, management incentive structures are typically biased towards short-term performance. A possible solution here is to link managerial compensation more closely to long-term, stock market earnings (Statman, 1982).

In larger organisations, departmental accountability for energy costs may be an important issue. If individual departments are accountable for their own energy costs, they could directly benefit from any savings from investment projects or housekeeping measures. But if cost savings are recouped elsewhere, this incentive is diluted. To introduce such accountability, it would be necessary to sub-meter and bill individual cost centres for their energy use – which would be associated with investment, staff and operational costs. The resulting incentives will be proportional to the importance of energy costs to the individual department and would only be effective if the department had the capacity to identify and initiate energy efficiency improvements. An alternative approach would be to place accountability for energy costs with the energy management staff, perhaps with individual posts made self-funding from the savings from energy efficiency improvements. The difficulty here is in identifying these savings and in adjusting for other sources of demand growth which are beyond the control of energy management staff. While energy management staff should have the capacity and skill to initiate energy efficiency improvements, they may lack local knowledge of individual efficiency opportunities. The appropriate solution will depend very much on the size and structure of the organisation and complex issues of accountability may arise.

Very similar issues arise in equipment purchasing. The purchaser may have a strong incentive to minimise capital costs, but may not be accountable for running costs. Many items of equipment may be specified and procured by individuals who lack the knowledge, information and incentives to minimise operating costs, while constraints on staff time may inhibit the involvement of energy management staff. In a similar manner, maintenance staff may have a strong incentive to minimise capital costs and/or to get failed equipment working again as soon as possible, but may have no incentive to minimise running costs. This type of issue may also arise with building users, operators of process equipment, and designers and sub-contractors within construction projects. In each case, the responsibility for capital costs may not match the responsibility for operating costs, while the transaction costs of reducing the problem may outweigh the potential savings to be achieved.

Within empirical research, the primary questions are: which types of split incentive problems are present within the case study organisations, and between these organisations and other market actors? What is their relative and aggregate importance in preventing improvements in energy efficiency? And what are the likely costs and benefits of changing the incentive structure?

5.6 Bounded Rationality

Several of the barriers discussed above (for example, risk) would still apply if actors were gifted with the perfect rationality assumed in orthodox theory. In other cases (for example, hidden costs), bounded rationality may be considered partly responsible for the barrier, since it contributes to the existence of transaction costs. But bounded rationality may also be classified as a barrier itself, since it contributes to decisions which depart from those predicted by orthodox models. This may occur even when actors have adequate motivation, incentives and information and when other barriers to energy efficiency are absent. For example, Eyre has argued that: '...There is a market failure to the extent that consumers do not attempt to maximise their utility or producers their profits' (Eyre, 1997, p. 36). The benchmark for this judgement is the 'optimising' rationality assumed in orthodox theory. As Sanstad and Howarth (1994b, p. 179) note: '...individuals and firms do not always behave according to the logic of economic rationality *but they should.* They need policies to help them do it.' The findings of behavioural economics may allow these departures from full rationality to be predicted and may also inform the design of policy interventions.

A primary consequence of bounded rationality is that constraints on time, attention, resources and the ability to process information lead to optimising analyses being replaced by imprecise routines and rules of thumb. In organisations, this could mean focusing on core activities, such as the primary production process, rather than peripheral issues such as energy use. Decision-making may also be divided up between specialists, and abstract global objectives may be replaced by tangible sub-goals whose achievement can be measured (Simon, 1979, p. 501).

Empirical studies of energy decisions generally support the hypothesis of bounded rationality (Sanstad and Howarth, 1994b). For example, Stern (1986, p. 205) finds that consumers hold information on household energy use which is '...not only incomplete, but systematically incorrect'. Similarly, the provision of accurate information on costs and benefits does not necessarily improve the quality of decision-making. In a survey of energy information programmes, Robinson (1991) concludes that '... it is clear that, with the exception of some labelling programmes, energy information programmes on their own have not to date resulted in significant energy savings'. The implication here is that not only may bounded rationality provide an additional barrier to energy efficiency, it may also undermine the effectiveness of certain types of policy interventions (Sanstad and Howarth, 1994b). If agents lack the time or capacity to use existing information, there is little point in providing more information. This point is important as it directly contradicts the orthodox argument that intervention should be directed at correcting information market failures, rather than imposing performance standards. In practice, standards may be more effective in some circumstances as they bypass the problem of bounded rationality (Sanstad and Howarth, 1994b, p. 118).

Routines as a response to bounded rationality

Within organisations, most decisions are likely to be the consequence of applying a set of rules and routines to a situation, rather than a systematic analysis of alternatives. As Stern (1984, p. 109) notes: 'Organisations generally solve problems and respond to environmental demands by applying existing routines rather than developing new ones.' The simple payback rule can be considered as one such routine (Stern, 1984, p. 109). While software packages allow rates of return to be calculated very easily, the payback rule may still have an advantage as it is simple and easy to communicate (DeCanio, 1994, p. 114). Capital budgeting procedures could be considered a second type of rule, used in delegating the authority to spend money. Typically, the primary concern when evaluating an investment opportunity is whether there is money in the budget, rather than the rate of return (Stern, 1984, p. 110). Expenditures that exceed the budget (that is, break the rule) require administrative approval, a potentially complex and lengthy process that discourages attempts to do so. Routines therefore facilitate information handling and minimise transaction costs, but can be inflexible.

A valuable example of the importance of routines is given by de Almeida's study of the French market for energy efficient motors (de Almeida, 1998). When small end-users had to buy motors in an emergency, the only parameters they considered were delivery time and price. The rule of thumb was to buy the same type and brand as the failed motor from the nearest retailer. Similarly, maintenance departments in large firms evaluated motors only in terms of maintenance costs and reliability, and ignored energy consumption. The split incentives barrier is partly responsible here, since maintenance departments are accountable for maintenance costs and process reliability, rather than energy costs. But de Almeida (1998) argues that this barrier is reinforced by the time constraints on maintenance staff and their limited capacity to process information. This leads to the development of organisational routines that simplify motor procurement by ignoring energy efficiency. In this area, as in many others, transaction costs, split incentives and bounded rationality reinforce one another. The motor example also demonstrates the importance of analysing specific technology decisions, since the relative importance of bounded rationality may be expected to vary with the type of decision – for example between emergency replacement, routine replacement and new requirement.

Another example of routines comes from the construction industry, where designers rely heavily on simplified and outdated (but cognitively efficient) rules of thumb for sizing heating, ventilation and air-conditioning (HVAC) equipment (Lovins, 1992). This frequently results in equipment which is oversized in relation to the load, and which runs inefficiently on part load. Building designers rely on similar rules of thumb for acceptable capital cost per unit floor area, with the result that the potential trade-offs between capital and operating costs are overlooked (Lovins, 1992). Similar examples are to be found throughout engineering design, where professionals working to intense time pressures rely heavily on standardised designs (Lovins and Lovins, 1997).

Other types of rules and routines which may impact on energy efficiency include: operating procedures (such as leaving equipment running or on standby); safety and maintenance procedures; relationships with particular suppliers; design criteria; specification and procurement procedures; equipment replacement routines and so on. These may either be formally specified in written procedures or embedded in organisational practices. Since routines are a means of allocating attention, energy efficiency opportunities will receive little attention if they do not form part of standard routines and operating procedures. And to the extent that consideration of energy efficiency entails transaction costs or requires additional cognitive effort, it may easily be squeezed out by other priorities.

Inertia and the status quo bias

Routines can be surprisingly persistent and entrenched. For example, Fawkes and Jacques (1987) observed that staff in a brewery preferred to use an inefficient design of pump because it was 'easy to clean':

> '...Only after extensive tests and persuasive efforts did the brewers admit that it was just as easy to clean the more efficient pump... the brewers exhibited an almost fanatical unwillingness to even consider change.' (Fawkes and Jacques, 1987, p. 564)

This type of problem has been labelled *inertia* within the energy efficiency literature and identified as a relevant explanatory variable for the efficiency gap (Stern, 1984; Katzev and Johnson, 1987; Lutzenhiser, 1993). From an orthodox perspective, inertia does not constitute a recognised market failure and provides no grounds for policy intervention. But from a behavioural economics perspective, inertia is exactly what we would expect. First, we have the observation that gains are treated differently from losses. This means that opportunity costs will be undervalued relative to out-of-pocket costs, and foregone gains will be considered to be less painful than perceived losses. In the case of energy efficiency, organisations will consider themselves 'endowed' with their existing buildings, equipment and energy bill (Hewett, 1998, p. 2.17). The potential savings in energy costs from energy efficiency improvements will be considered an opportunity cost, while the investment costs of energy efficient equipment will be considered an out-of-pocket cost. Loss aversion will therefore tend to bias individuals against making such improvements. As Thaler notes: '[A] certain degree of inertia is introduced into the consumer choice process since goods that are included in the individual's endowment will be more highly valued than those not held in the endowment ...' (Thaler, 1991, p. 8).

Second, we have the observation that individuals tend to be risk-averse with respect to gains. In the case of energy efficiency investment, uncertainty over factors such as technology performance, reliability, lifetime and length of ownership will create uncertainty in the potential energy savings. In contrast, continuing with the existing 'endowment' is likely to give more predictable outcomes (Hewett, 1998, p. 2.17). Since outcomes that are known with certainty will be given greater weighting than those that are uncertain, this will reinforce the tendency to inertia.

A third factor is the desire to minimise regret (Hewett, 1998, p. 2.17):

> 'Action and decisions require a greater justification than inaction, than failing to decide…If our actions do not pan out, or cause a loss, we regret having acted. If, instead, we do not act, if we leave things as they are, and our investment does not pan out, or we lose, we still suffer regret but the regret is *lesser*.' (Piattelli-Palmarini, 1994, pp. 27–8)

All these factors may cause individuals and organisations to favour the status quo and to neglect potential improvements in energy efficiency, even when other market and organisational failures are absent. It is in this sense that cognitive biases may be considered as an additional barrier to energy efficiency. Rather similar conclusions were drawn by Stern (1994) in a review of behavioural research on energy efficiency. Stern utilised a theory known as 'cognitive dissonance' to conclude that: a) people tend to rationalise previous decisions, emphasising the positive aspects of the decision and the negative aspects of the unchosen alternative; b) this tendency is greater for difficult, costly or irreversible decisions; and c) people remember the plausible arguments for their own position and forget the plausible arguments opposing their position. Hence people resist change because they are committed to what they are doing, and they justify that inertia by the downgrading of contrary information.

As Williamson (1988) argues, one of the functions of organisations is to economise on bounded rationality and mitigate such biases through the use of specialisation. But while biases may be reduced, they are unlikely to be eliminated. Similarly, while competitive market pressures should help to squeeze out inefficiencies within private firms, the potential savings through energy efficiency may be relatively small compared to the other determinants of competitive advantage and hence may persist.[20] In practice, empirical studies of managerial decision-making have suggested that loss aversion and risk aversion may be even stronger than in the individual context (Swalm, 1966). Kahneman and Lovallo (1993) note that decision-makers become more risk-averse when they expect choices to be reviewed by others, while the rhetoric of prudent decision-making favours the certainty effect. For example, Swalm (1966) cites a manager who declined to pursue a project that had a 50-50 chance of either making $300k or losing $60k.

The above observations are generic – inertia may prevent the take-up of a wide range of opportunities, not simply those related to energy efficiency. But the important point is the effect of inertia on energy efficiency investment *relative* to that on purchasing energy as a commodity. This is analogous to the effect of imperfect information on energy efficiency investment relative to the purchase of energy commodities. In both cases, we would expect there to be a bias against energy efficiency and in favour of purchasing energy commodities. Inertia matters more for energy efficiency because it involves investing in hardware with uncertain outcomes and because it represents a departure from the status quo. Since energy efficiency and energy purchasing can provide alternative means of delivering the same energy service, this may be economically inefficient.

6. SUMMARY AND CONCLUSIONS

This chapter has attempted to improve the understanding of barriers to energy efficiency by applying a number of concepts from economic theory. A 'barrier' was defined as a mechanism that inhibits a decision or behaviour that appears to be both energy efficient and economically efficient. This term is widely used within the energy efficiency literature, but there is no consensus on how barriers should be understood, how important they are in different contexts, and how (if at all) they should be addressed. This makes barriers the subject of disciplinary disputes within academia and more fundamental conflicts within the politics of climate change.

The chapter began by introducing the concept of a barrier, the history of the barriers debate and the criticisms of the barrier model by orthodox economists. This led to a distinction between barriers and orthodox market failures and a recognition that some barriers may provide no grounds for policy intervention while others may prove too costly to overcome. The orthodox perspective was criticised, however, for treating market failures as absolute rather than relative, for ignoring barriers which were internal to organisations and for adopting an unrealistic model of individual rationality.

The chapter then introduced a series of ideas from orthodox, transaction cost and behavioural economics which were argued to provide useful insights into the barriers debate. These included the informational problems associated with different types of good, the origin, nature and consequences of adverse selection and moral hazard, the problem of split incentives and appropriability, the nature and determinants of transaction costs, and the consequences of bounded rationality and 'cognitive illusions'. Each of these is the subject of a considerable theoretical and empirical literature and each may be used to interpret the observations of energy efficiency practitioners within a formal theoretical framework. However, there is a great deal of overlap between these perspectives, and the barriers described in the energy efficiency literature may frequently subsume more than one of these ideas and refer to phenomena that may be explained by more than one mechanism.

The chapter then introduced a taxonomy of six barriers to energy efficiency, using terminology that is widely employed within the energy efficiency literature. The nature of these barriers was then interpreted using the more formal economic concepts, and the operation and consequences of these barriers explored in some detail. The chapter argued that each barrier may be explained by more than one of the ideas discussed earlier and that several mechanisms may be expected to coexist. For example, hidden costs may result from search costs within product markets, organisational transaction costs and the costs resulting from interruptions to production. Similarly, restrictions on capital budgets may result from the financing risk of increased gearing, the agency relationships between central and divisional management, and the transaction costs of raising additional internal or external funds. The relative importance of each factor may be expected to vary between different technologies and organisations.

Table 2.8 summarises how the proposed barriers may be understood using these more formal economic ideas. Each of the barriers is subdivided into one or more mechanisms that may potentially explain the neglect of individual

Table 2.8 Understanding barriers to energy efficiency from orthodox, agency, transaction cost and behavioural perspectives

Barrier	**Orthodox and agency perspectives**	**Transaction cost and behavioural perspectives**
Risk	• Energy efficient technologies may be subject to higher technical risk • Energy efficiency investments may be illiquid and irreversible, so the value of future benefits should be discounted more highly • Energy efficiency investments may require a high rate of return to reflect the option value of delaying investment	• Energy efficiency investments may have a high asset specificity and hence carry higher risks than other forms of investment • Individual and organisational perceptions of risk may depart from those assumed in orthodox models and the resulting loss aversion may create a bias against energy efficient investment
Imperfect information	• Some types of information relevant to energy efficiency may have the features of a public good and hence may be undersupplied by markets. • Since energy efficiency has the characteristics of a credence good, while energy commodities have the characteristics of a search good, there may be a systematic bias against the former. • The search costs associated with identifying the energy consumption of products may be high, thereby creating a barrier to such purchases. • In some circumstances, asymmetric information within energy service markets may lead to the adverse selection of inefficient products.	• There will be transaction costs associated with acquiring, understanding and applying information. Bounded rationality ensures that these will be significant, even when information is freely available • The transaction costs of information acquisition may be high as a consequence of poor presentation, the lack of credibility of the source or the absence of interpersonal contacts

Hidden costs	• Some energy efficient options may be associated with hidden costs, such as disruptions to production • Some energy efficient technologies may perform poorly along other dimensions, such as reliability • The search costs associated with identifying energy efficient products may be high • Energy efficiency investments may need to recover 'overhead' costs	• The transaction costs associated with maintaining information systems, conducting energy audits, identifying opportunities, tendering, selecting suppliers, seeking approval for capital expenditures and so on may outweigh the potential savings in energy costs
Access to capital	• The cost of obtaining additional capital may exceed the average cost of the existing debt/equity mix, owing to the financing risk of increased gearing • Agency problems between shareholders and managers, plus the signalling associated with external financing may create a preference for internal finance and a reluctance to increase borrowing • Agency problems between central and divisional managers may lead to capital rationing as a form of control • There may be asymmetry between the risks and rewards of energy efficiency projects.	• There will be transaction costs associated with obtaining additional funding from either internal or external sources, and these may be more significant for small, cost saving projects • The transaction costs of transmitting and assessing information on investment opportunities may inhibit optimal decision-making • Given severe constraints on time and attention, managers may focus on strategic investments and overlook small cost saving opportunities

Barrier	Orthodox and agency perspectives	Transaction cost and behavioural perspectives
Split incentives	• For a range of reasons, individuals, departments or organisations may not be able to appropriate the benefits of energy efficiency investments	• Transaction costs inhibit the development of shared savings contracts to overcome split incentive problems
Bounded rationality	• N/A	• Constraints upon time, attention, resources and ability to process information lead to the use of imprecise routines and rules of thumb, which may systematically neglect the small cost savings from energy efficiency improvements • Loss aversion and status quo bias contributes to inertia and the undervaluing of the benefits of energy cost saving relative to the out-of-pocket costs of investment • Risk aversion with respect to gains reinforces this inertia

energy efficiency opportunities. The case study research described in the following chapters will use this framework to interpret the empirical results.

NOTES

1. Williamson (1988) provides a useful comparison of agency theory and transaction cost economics.
2. Public intervention may also be justified for a number of other reasons, such as improved equity or democratic accountability, but this is unlikely to improve economic efficiency.
3. Jaffe and Stavins note that: 'Some analysts have identified the transaction costs associated with adopting new technologies (such as costs associated with searching and learning) as an additional category of market failure...Other than with regard to the public good dimensions of learning, such adoption costs are more aptly described as part of the typically unmeasured costs of technology and hence, if anything, constitute a category of non-market failure explanations of the paradox...' (Jaffe and Stavins, 1994, p. 809).
4. For a valuable illustration of this, see Howarth et al. (2000).
5. Defined as the sum of consumer surplus (the integral of the difference between consumer valuations of a good and its price) and producer surplus (the integral of the difference between sales revenue and marginal production costs).
6. Rivalry is primarily determined by the physical characteristics of a good. For example, no one else can wear my coat if I am wearing it. In contrast, excludability is determined by both the physical characteristics of a good and the legal rules defining property rights. For example, it is physically possible for my neighbour to share my lawn, but legal rules may prevent her doing so without my consent. In some cases, property rights may be partitioned, such as when the right to use a car is transferred to another through a rental contract. But here, the property rights remain well-defined and enforceable.
7. A market failure may be said to exist when the producer does not supply the amount of information that maximises the difference between the reduction in the associated efficiency loss and the cost of providing the information (Weimer and Vining, 1999, p. 70).
8. While this represents a somewhat jaundiced view of human nature, it is not necessary to assume that all parties are given to opportunism in order to recognise the value of this assumption in explaining a number of observed economic arrangements and outcomes. For example, it is not necessary to assume that all people are potential thieves in order to explain the commonly observed practice of locking a house.
9. Implicitly, Akerlof's model assumes imperfect insurance markets (seller could provide warranty) or high costs of bringing the car to a garage to have it checked.
10. For example, loan applicants may have private information about their creditworthiness or business prospects. This information is not accessible by lenders, and the applicants have no incentive to disclose it. Raising the interest rate charged to borrowers may only lead to a poorer quality of loan applicants, which in turn may raise the bank's costs of lending (Stiglitz and Weiss, 1981).
11. For their work on asymmetric information, Akerlof, Spence and Stiglitz were rewarded the Nobel Price for economics in 2001. Akerlof also said: 'Economic theorists, like French chefs in regard to food, have developed stylised models whose ingredients are limited by some unwritten rules. Just as traditional French cooking does not use seaweed or raw fish, so neoclassical models do not make assumptions derived from psychology, anthropology, or sociology. I disagree with any rules that limit the nature of the ingredients in economic models.'
12. 'Most human decision making, whether individual or organisational, is concerned with the discovery and selection of satisfactory alternatives; only in exceptional cases is it concerned with the discovery and selection of optimal alternatives. To optimise requires

processes several orders of magnitude more complex than those required to satisfice.' (March and Simon, 1958, p. 171).

13. In the textbook model of competitive general equilibrium, all contracts are assumed to be complete in that they specify a course of action, decision or terms of trade contingent on every possible future state of affairs.
14. Douglas North (1990) has pioneered the application of transaction costs to large-scale institutions and economic history, while Williamson has focused primarily on microeconomics.
15. The problem is well expressed by Foss (2001): 'It is because the ocean of bounded rationality is so vast and so deep that economists have preferred to stay to the low waters close to the continent of full rationality, lest they fall victim to drowning accidents. Even better, avoid the waters entirely' (Foss, 2001, p. 6).
16. Kahneman and Tversky (1979) have shown how individuals will chose A over B if the choice is framed in one way, and B over A if the choice is framed in a different way. This violates the invariance requirement of rational decision-making.
17. In the UK, the liberalisation of gas and electricity supply markets has increased competition and choice while at the same time increasing search costs. This process has undoubtedly brought benefits to large users, but the benefits for household consumers are less clear (Waddams, 2003).
18. For example, a US study showed that people who viewed a video about implementing energy saving measures were significantly more likely to cut energy use than those who received the information in writing (Winett et al.., 1982, p. 24).
19. This is calculated from the relative proportions of loan stock and equity for the individual company, and their respective market values (Mclaney, 2000, p. 249).
20. It is an empirical question as to whether the potential savings are really small, or are merely perceived to be small. Lovins and Lovins (1997, p. 11) quote the example of a chief executive officer of a Fortune 100 company who stated that 'I can't really get excited about energy – it's only a few percent of my cost of doing business.' But if the cost savings achieved by the energy manager at one of his company's sites could have been reproduced throughout the company, it would have boosted the net earnings that year by 56 per cent.

REFERENCES

Akerlof, G. (1970), 'The market for lemons', *Quarterly Journal of Economics*, **84**(3), 488–500.

Allen, D.W. (1996), 'Transaction costs', *Encyclopaedia of Law and Economics*, Edward Elgar, Cheltenham.

Antle, R. and G.D. Eppen (1985), 'Capital rationing and organisational slack in capital budgeting', *Management Science*, **31**(2), 163–74.

Arrow, K. (1969), 'The organisation of economic activity: issues pertinent to the choice of market versus non-market allocation', *The Analysis and Evaluation of Public Expenditure: The PPB system: Vol. 1*, US Joint Economic Committee, 91st Congress, 1st Session, Washington DC, US Government Printing Office, 59–73.

Asquith, P. and D.W. Mullins (1986), 'Equity issues and offering dilution', *Journal of Financial Economics*, **15**: 61–89.

Bromley, D.W. (1991), *Environment and Economy: Property Rights and Public Policy*, Blackwell, Oxford.

Camerer, C. (1998), 'Bounded rationality in individual decision-making', *Experimental Economics*, **1**, 163–83.

Conlisk, J. (1996), 'Why bounded rationality?' *Journal of Economic Literature*, **XXXIV**, 669–700.

Craig, C.S. and J.M. McCann (1978), 'Assessing communication effects on energy conservation', *Journal of Consumer Research*, **5**, 82–8.

de Almeida, E.L.F. (1998), 'Energy efficiency and the limits of market forces: the example of the electric motor market in France', *Energy Policy*, **26**(8), 643–53.

DeCanio, S.J. (1993), 'Barriers within firms to energy efficient investments', *Energy Policy*, **21**(9), 906–14.

DeCanio, S.J. (1994), 'Agency as control problems in US corporations: the case of energy efficient investment projects', *Journal of the Economics of Business*, **1**(1), 105–23.

Department of Energy (1983), 'Energy conservation investment in industry: an appraisal of the opportunities and barriers', Armitage Norton Consultants, Energy Paper No. 50, HMSO, London.

Dixit, A.K. and R.S. Pindyck (1994), *Investment under Uncertainty*, Princeton University Press, Princeton.

Douma, S. and H. Schreuder (1998), *Economic Approaches to Organisations*, Prentice Hall, London.

Energy Efficiency Office (1993), 'Organisational aspects of energy management', General Information Report 12, ETSU, Harwell.

Ester, P. and W.R.A. Winett (1982), 'Towards more effective antecedent strategies for environmental programmes', *Journal of Environmental Systems*, **11**, 201–21.

ETSU (Energy Technology Support Unit) (2001), *Climate Change Agreements – Sectoral Energy Efficiency Targets*, ETSU, AEA Technology, Harwell.

Eyre, N. (1997), 'Barriers to energy efficiency: more than just market failure', *Energy and Environment*, **8**(1), 25–43.

Fawkes, S.D. and J.K. Jacques (1987), 'Approaches to energy conservation management in beverage related industries and their effectiveness', *Energy Policy*, **15**(6), 559–66.

Foss, N.J. (2001), 'From thick to think bounded rationality in the economics of organisation: an explorative discussion', Working Paper, Department of Industrial Economics and Strategy, Copenhagen Business School, Copenhagen.

Foss, N.J. (2003), 'Bounded rationality in the economics of organisation: "Much cited and little used"', *Journal of Economic Psychology*, **24**, 245–64.

Furubotn, E.G. and R. Richter (1997), *Institutions and Economic Theory: The Contribution of the New Institutional Economics*, University of Michigan Press, Ann Arbor.

Goldstone, S. (1995), 'Restructuring: stimulus to improving utility DSM: how economists might help', Western Economic Association 70th annual conference, San Diego, July 5–9.

Golove, W.H and J.H. Eto (1996) 'Market barriers to energy efficiency: a critical reappraisal of the rationale for public policies to promote energy efficiency', LBL-38059, Lawrence Berkeley Laboratory, University of California, Berkeley.

Haddad, B.M., R.B. Howarth and B. Paton (1998), 'Energy efficiency and the theory of the firm', Proceedings from the 1998 ACEEE Summer Study on Energy Efficiency in Buildings, American Council for an Energy Efficient Economy, Washington DC.

Hassett, K.A. and G.E. Metcalf (1993), 'Energy conservation investment: do consumers discount the future correctly?', *Energy Policy*, **23**(1), 710–16.

Hein, L. and K. Blok (1994), 'Transaction costs of energy efficiency improvement', Proceedings, European Council for an Energy-Efficient Economy.

Herring, H., R. Hardcastle and R. Philipson (1988), 'Energy use and energy efficiency in UK commercial and public buildings up to the year 2000', Energy Efficiency Office, Department of Energy, HMSO, London.

Hewett, M.J. (1998), 'Achieving energy efficiency in a restructured electric utility industry', Report prepared for Minnesotans for an Energy Efficient Economy, Center for Energy and Environment, Minneapolis, MN.

Hirst, E. and M. Brown, (1990), 'Closing the efficiency gap: barriers to the efficient use of energy', *Resources, Conservation and Recycling*, **3**, 267–81.

Hodgson, G.M. (1988), *Economics and Institutions: A Manifesto for a Modern Institutional Economics*, Polity Press, Cambridge.

Holmstrom, B. (1979), 'Moral hazard and observability', *Bell Journal of Economics*, **19**, 74–91.

Howarth, R.B. and A.H. Sanstad (1995), 'Discount rates and energy efficiency', *Contemporary Economic Policy*, **XIII**, 101–9.

Howarth, R.B., B.M. Haddad and B. Paton (2000), 'The economics of energy efficiency: insights from voluntary participation programmes', *Energy Policy*, **28**(6-7), 477–86.

Huntington, H.G., L. Schipper and A.H. Sanstad (1994), 'Editor's introduction', *Energy Policy*, **22**(10), 795–97.

Jaffe, A.B. and R.N. Stavins (1994), 'The energy-efficiency gap: what does it mean?', *Energy Policy*, **22**(10), 804–10.

Jensen, M.C. (1986), 'Agency cost of free cash flow: corporate finance and takeovers', *American Economic Review*, **76**, 323–29.

Johnson, B.E. (1994), Modelling energy technology choices: which investment tools are appropriate?, *Energy Policy*, **22**(10), 877–83.

Joskow, P.L. (1985), 'Vertical integration and long-term contracts: the case of coal-burning electrical generating plants', *Journal of Law, Economics and Organisation*, **1**, 33–80.

Kahneman, D. and A. Tversky (1979), 'Prospect theory: an analysis of decision making under risk', *Econometrica*, **47**, 263–91.

Kahneman, D. and A. Tversky (2000), *Choices, Values and Frames*, Cambridge University Press, Cambridge.

Kahneman, D. and D. Lovallo (1993), 'Timid choices and bold forecasts: a cognitive perspective on risk taking', *Management Science*, **39**(1), 17–31.

Kahneman, D., J.K. Knetsch and R.H. Thaler (1991), 'The endowment effect, loss aversion and status quo bias', *Journal of Economic Perspectives*, **5**(1), 193–206.

Katzev, R. and T. Johnson (1987), *Promoting Energy Conservation: An Analysis of Behavioural Research*, Westview Press, Boulder.

Kempton, W. and L. Montgomery (1982), 'Folk quantification of energy', *Energy*, **7**(10), 817–27.

Levine, M.D., E. Hirst, J. Koomey, J. McMahon and A. Sanstad (1994), 'Energy efficiency, market failures and government policy', Lawrence Berkeley Laboratory and Oak Ridge Laboratory, LBL 35376, ORNL/CON-383.

Lovins, A. (1992), *Energy Efficient Buildings: Institutional Barriers and Opportunities*, Strategic Issues Paper No. 1, E Source Inc., Boulder Colorado.

Lovins, A.B. and L.H. Lovins (1997), *Climate: Making Sense and Making Money*, Rocky Mountain institute, Colorado.

Lutzenhiser, L. (1993), 'Social and behavioural aspects of energy use', *Annual Review of Energy and the Environment*, **18**, 247–89.
March, J. G. and H. Simon (1958), *Organisations*, John Wiley and Sons, New York.
McDonald, R. and D. Siegel (1986), 'The value of waiting to invest', *Quarterly Journal of Economics*, **101**(4), 707–27.
Mclaney, E.J. (1994), *Business Finance for Decision Makers*, 2nd Edition, Pitman Publishing, London.
Metcalf, G.E. (1994), 'Economics and rational conservation policy', *Energy Policy*, **22**(10) 819–25.
Milgrom, P. and J. Roberts (1992), *Economics, Organisation and Management*, Prentice Hall, New Jersey.
Modigliani, F. and M.H. Miller (1958), 'The cost of capital, corporate finance and the theory of investment', *American Economic Review*, **48**, 261–97.
Mohlo, I. (1998), *The Economics of Information: Lying and Cheating in Markets and Organisations*, Blackwell, London.
Myers, S.C. (2001), 'Capital structure', *Journal of Economic Perspectives*, **15**(2), 81–102.
Myers, S.C. and N.S. Majluf (1984), 'Corporate financing and investment decisions when firms have information that investors do not have', *Journal of Financial Economics*, **13**, 187–221.
Narayanan, M.P. (1985), 'Observability and the payback criterion', *Journal of Business*, **98**(3), 309–23.
Nelson, R. and S. Winter (1982), *An Evolutionary Theory of Economic Change*, Belknap Press, Cambridge, MA.
Nichols, A.L. (1994), 'Demand-side management: overcoming market barriers or obscuring real costs?', *Energy Policy*, **22**(10), 840–7.
North, D. (1990), *Institutions, Institutional Change and Economic Performance*, Cambridge University Press, Cambridge.
OFT (Office of Fair Trading) (1997), 'Consumer detriment under conditions of imperfect information', Research Paper 11, London.
Ostertag, K. (1999), 'Transaction costs of raising energy efficiency', paper presented at the IEA International Workshop on Technologies to Reduce Greenhouse Gas Emissions: Engineering-Economic Analyses of Conserved Energy and Carbon, 5–7 May 1999, Washington DC.
Ostertag, K. (2003), *No-Regret Potentials in Energy Conservation: An Analysis of their Relevance, Size and Determinants*, Physica: Heidelberg.
Piattelli-Palmarini, M. (1994), *Inevitable Illusions: How Mistakes of Reason Rule Our Minds*, John Wiley and Sons, New York.
Pindyck, R.S. and D.L. Rubenfeld (1998), *Microeconomics*, Prentice Hall, Upper Saddle River.
Prendergast, C. (1999), 'The provision of incentives in firms', *Journal of Economic Literature*, **XXXVII**, 7–63.
Rabin, M. (1998), 'Psychology and economics', *Journal of Economic Literature*, **36**(1), 11–46.
Rabin, M. (2002), 'A perspective on psychology and economics', *European Economic Review*, **46**, 657–85.
Robinson, J. B. (1991), 'The proof of the pudding: making energy efficiency work', *Energy Policy*, **19**(7), 631–45.
Ross, M. (1986), 'The capital budgeting practices of 12 large manufacturing firms', *Financial Management*, **15**(4), 15–22.

Samuelson, W. and R. Zeckhauser (1988), 'Status quo bias in decision-making', *Journal of Risk and Uncertainty*, **1**(1), 7–59.

Sanstad, A.H. and R.B. Howarth (1994a), '"Normal" markets, market imperfections and energy efficiency', *Energy Policy*, **22**(10), 811–18.

Sanstad, A.H. and R.B. Howarth (1994b), 'Consumer rationality and energy efficiency', ACEEE Summer Study on Energy Efficiency in Buildings, ACEEE.

Sanstad, A.H., C. Blumstein and S.E. Stoft (1995), 'How high are option values in energy efficiency investments?', *Energy Policy*, **23**(9), 739–43.

Schleifer, A. and R.W. Vishny (1997), 'A survey of corporate governance', *Journal of Finance*, **LII**(2), 737–83.

Seligman, C., L.J. Becker, and L.J. Darley (1981), 'Encouraging residential energy conservation through feedback', in A. Baum and J.E. Singer (eds), *Advances in Environmental Psychology: Vol 3: Energy: Psychological perspectives*, Hillsdale, NJ: Erlbaum, 93–113.

Shelanski, H.A. and P.G. Klein (1995), 'Empirical research in transaction cost economics: a review and assessment', *Journal of Law, Economics and Organisation*, **11**(2), 335–61.

Simon, H.A. (1957), *Models of Man*, Wiley, London.

Simon, H.A. (1959), 'Theories of decision-making in economics and behavioural science', *American Economic Review*, **XLIX**(3), 253–83.

Simon, H.A. (1979), 'Rational decision making in business organizations', *American Economic Review*, **69**, 493–513.

Simon, H.A. (1983), *Reason in Human Affairs*, Stanford University Press, Stanford.

Simon, H.A. (1986), 'Rationality in psychology and economics', *Journal of Business*, **59**, 209–24.

Sorrell, S. (1996), 'The economics of industrial and commercial CHP: lessons from the UK', a report to the Co-operative Programme on Environment and Development, SPRU – Science and Technology Policy Research, University of Sussex.

Sorrell, S. (2002), 'The climate confusion: implications of the EU emissions trading scheme for the UK Climate Change Levy and Climate Change Agreements', SPRU – Science and Technology Policy, University Of Sussex, Brighton.

Spence, M. (1973), 'Job market signalling', *Quarterly Journal of Economics*, **87**, 355–74.

Statman, M. (1982), 'The persistence of the payback permit: a principal-agent perspective', *Engineering Economist*, **27**(2), 95–100.

Statman, M. and J.F. Sepe (1984), 'Managerial incentive plans and the use of the payback method', *Journal of Business Finance and Accounting*, **11**(1), 61–5.

Stein, J.C. (2001), 'Agency, information and corporate investment', working paper 8342, National Bureau of Economic Research, Cambridge, Massachusetts.

Stern, P.C. (1992), 'What psychology knows about energy conservation', *American Psychologist*, **47**, 1224–32.

Stern, P.C. (1984), *Energy Use: The Human Dimension*, W.H. Freeman, New York.

Stern, P.C. (1986), 'Blind spots in policy analysis: what economics doesn't say about energy use', *Journal of Policy Analysis and Management*, **5**(2), 200–227.

Stigler, G.J. (1961), 'The economics of information', *Journal of Political Economy*, **69**(3), 213–25.

Stiglitz, J.E. (1975), 'The theory of screening: education and the distribution of income', *American Economic Review*, **65**(3), 283–300.

Stiglitz, J.E. and A. Weiss, (1981), 'Credit rationing in market with imperfect information', *American Economic Review*, **71**, 393–409.

Sutherland, R.J. (1991), 'Market barriers to energy efficiency investments', *Energy Journal*, **12**(3), 15–34.

Sutherland, R.J. (1996), 'The economics of energy conservation policy', *Energy Policy*, **24**(4), 361–70.

Swalm, R.O. (1966), 'Utility theory: insights into risk-taking', *Harvard Business Review*, **44**, 39–60.

Thaler, R.H. (1980), 'Towards a positive theory of consumer choice', *Journal of Economic Behaviour and Organisation*, **1**, 39–60.

Thaler, R.H. (1991), *Quasi Rational Economics*, New York, Russell Sage Foundation.

Thaler, R. H. (2000), 'From Homo Economicus to Homo Sapien', *Journal of Economic Perspectives*, **14**(1), 133–41.

Train, K. (1985), 'Discount rates in consumers' energy related decisions: a review of the literature, *Energy*, **10**(12), 1243–53.

Waddams, C. (2003), 'Spoilt for choice? The costs and benefits of opening UK residential energy markets', paper presented at the 2003 British Institute of Energy Economics Academic Conference, St John's College, Oxford, 25–26 September.

Wade, J., J. Pett and L. Ramsay (2003), *Energy Efficiency in Offices: Assessing the Situation*, Association for the Conservation of Energy, London.

Warren, A. (1987), 'Saving megabucks by saving megawatts', *Energy Policy*, December, **15**(6), 522–37.

Weber, L. (1997), 'Some reflections on barriers to the efficient use of energy', *Energy Policy*, **25**(10), 833–5.

Weimer, D.L. and A.R. Vining (1999), *Policy Analysis: Concepts and Practice*, Prentice Hall, New Jersey.

Williamson, O. E. (1979), 'Transaction cost economics: the governance of contractual relations', *Journal of Law and Economics*, **22**, 233–61.

Williamson, O.E. (1985), *The Economic Institutions of Capitalism*, Free Press, New York.

Williamson, O.E. (1988), 'Corporate finance and corporate governance', *Journal of Finance*, **XLIII**(3), 567–91.

Williamson, O.E. (1989), 'Transaction cost economics', in R. Schmalensee and R.D. Willig (eds),*Handbook of Industrial Organisation: Volume 1*, Elsevier, North-Holland Amsterdam.

Winett, R.A., I.N. Leckliter, D.E. Chinn and B. Stahl (1984), 'Reducing energy-consumption – the long-term effects of a single TV program', *Journal of Communication*, **34**(3), 37–51.

3. Energy Costs – Why Bother? Barriers to Energy Efficiency in the German Higher Education Sector

Joachim Schleich

'...it is not important to save energy or energy costs, but rather to comply with regulations'.[1]

1. INTRODUCTION

Total energy costs for the German higher education (HE) sector are approximately €500 million/year, but individual institutions only spend around 2 per cent of their budget on energy. As a consequence, little attention is paid to energy efficiency in university decision-making which is dominated by research and teaching concerns. Nevertheless, total energy consumption by the HE sector is quite significant, accounting for about 0.4 per cent of German final energy consumption. The majority of this is for generic uses, notably heating, ventilation, air conditioning (HVAC) and lighting. Hence, unlike the brewing sector but similar to the mechanical engineering sector, energy-conversion technologies in HE are cross-cutting rather than process-specific.

The German HE sector is being increasingly affected by severe restrictions on federal and state expenditure, with the result that investments in new equipment and buildings are being postponed or shelved. From a national policy point of view this poses a particular challenge, since Germany – for lack of other resources – has relied extensively on a strong educational sector in the past. For these and other reasons, the HE sector is undergoing major reforms which will also affect universities' attitudes towards energy management and their ability to overcome barriers to energy efficiency.

The research for this study was carried out in 1999, about one year after the liberalisation of the German electricity market. Liberalisation has had two important consequences. First, and most visibly, there was a dramatic drop in electricity prices for wholesale and end-users.[2] Large industrial customers were the first to benefit from price decreases of up to 40–50 per cent and, since the summer of 1999, private households have also enjoyed price cuts of up to 30 per cent. These price cuts have rendered investments in energy efficient measures less profitable in all end-use sectors, including HE. Secondly, in the wake of liberalisation, a large number of energy service companies (ESCOs) emerged, many of which focused on the public sector market. On the one hand, these ESCOs may realise energy savings potentials that would otherwise have been left untapped. On the other hand, there is some opposition to energy service concepts from those traditionally in charge of energy management at public sector institutions.

This chapter explores barriers to energy efficiency in the German HE sector and offers some recommendations on how these could be overcome. The study is based on six in-depth case studies of individual universities and a number of additional interviews with ESCOs. The chapter first provides an overview of the sector, summarises the pattern of energy use and energy costs and describes both the determinants of energy performance and the decision-making procedure for investments. The characteristics and management practices of the case study universities are then described, followed by an assessment of the relative importance of different barriers to energy efficiency. Recommendations are then presented on how these barriers may be overcome and how energy efficiency in the sector may be improved. In particular, the limiting and success factors for ESCOs in the sector are identified and assessed.

2. THE GERMAN HIGHER EDUCATION SECTOR

2.1 Sector Overview

During the winter semester of 1998/99, there were 349 HE institutions in Germany, of which 86 were universities.[3] Since the early 1970s the total number of HE students has grown from 422,000 to over 1.8 million (Statistisches Bundesamt, 1999a,b). During the winter semester 1998/1999, 481,000 people were employed in the HE sector, of which 45 per cent were scientific staff and 37,600 were professors. Operating expenses for HE institutions are largely financed through the budgets of the federal states (*Laender*), with only a small contribution from churches or private organisations. Investment costs for large equipment, new buildings and building extensions are evenly split between the state and federal budgets. The total expenditure for all HE institutions by the federal and state budgets amounted to almost €26,000 million in 1997, of which €3,000 million (11.7 per cent) was for capital investments.

Like almost all publicly-funded projects and institutions, higher education in Germany suffers from tight federal and state budgets. Since many institutions need new equipment for research and teaching and many buildings need to be refurbished, the budget situation constitutes a major challenge for the proper functioning of the HE sector (Planungsausschuss, 2000). Difficulties also arise from current reforms in the HE sector, including increased autonomy from the state authorities, the introduction of global budgeting and business accounting, new funding schemes, tuition payments in some states, rankings and evaluation, increased competition from other public or private universities, new forms of teaching such as virtual universities and new designs for bachelors' and masters' degrees. But as described below, some of these reforms may also be beneficial for energy efficiency.

2.2 Energy Consumption and Energy Efficiency

There is no data available on the total or specific energy consumption in the German HE sector, but total final energy consumption can be estimated using data on specific energy consumption from a subset of institutions (Schleich and Boede, 2000)[4]. According to these calculations, energy consumption in the German HE sector is approximately 40 PJ/year, which corresponds to 0.4 per cent of German final energy demand.[5] About 40 per cent of energy consumption is for electricity use and 60 per cent for space or water heating. Electricity is used for ventilation and air conditioning (30–50 per cent), lighting (20–30 per cent) and office equipment (20–30 per cent), and accounts for around 60 per cent of total energy costs. Energy costs for the entire HE sector are approximately €500 million/year, or 2 per cent of total annual expenditure by the HE sector.

The technical and economic potential for energy efficiency in the German HE sector is estimated to be substantial. Government analyses have estimated that organisational measures in the public sector may save 5–15 per cent, while technical measures could reduce thermal energy consumption by 25–60 per cent and electricity consumption by at least 10 per cent (Umweltbundesamt, 1999). Similarly, a recent study by the Austrian Energy Agency for Austrian universities (Energieverwertungsagentur, 1999), suggested that information and education measures that changed users' behaviour could achieve 20–30 per cent energy savings without loss of comfort and at low to zero capital costs.[6]

Energy efficiency opportunities in the HE sector are typically cross-cutting. The measures selected for investigation in the case study institutions are listed in Table 3.1. These represent established and low-risk technologies which are widely recommended in the best-practice literature.

Table 3.1 Selected technical measures for energy efficiency in German universities

Area	**Technology/technique**
Heating	Use of Building Energy Management System (BEMS)
	Programming heating and ventilation controls to match occupancy and/or temperature
	Ensuring thermostats and temperature sensors are located in the correct place
	Installation of thermostatic radiator valves
	Improvements to the zoning of heating areas
	Restricting use of portable electric heaters
Lighting	Use of compact fluorescents
	Replacement of 38mm fluorescents with 26mm
	Use of high frequency electronic ballasts
	Use of photocell, acoustic or movement sensors
	Lighting controls integrated into BEMS
Plant	Use of boiler sequencing controls
	Insulation of pipes and valves
	Replacement of oversized boiler plant
	Replacement of central generation of hot water with point of use applications
	Installation of condensing boilers
	Installation of CHP
Fabric	Fitting door closers to external doors
	Installation of secondary or double glazing on refurbishment
	Specification of insulation standards in new buildings that exceed the building regulations
	Retrofitting insulation to walls and roofs
	Draught proofing of windows and doors
Electrical	Power factor correction
	Use of variable speed drives (VSD) in pumps, fans and other applications
	Automatic switch off of fans and pumps
	Specification of high efficiency motors on motor replacement
	Specification of high efficiency office equipment
	Specification of high efficiency catering equipment

2.3 Determinants of Energy Performance

Figure 3.1 illustrates the institutions, groups and individuals that influence energy consumption and energy efficiency in German universities. The relevant actors are both internal and external to the university.

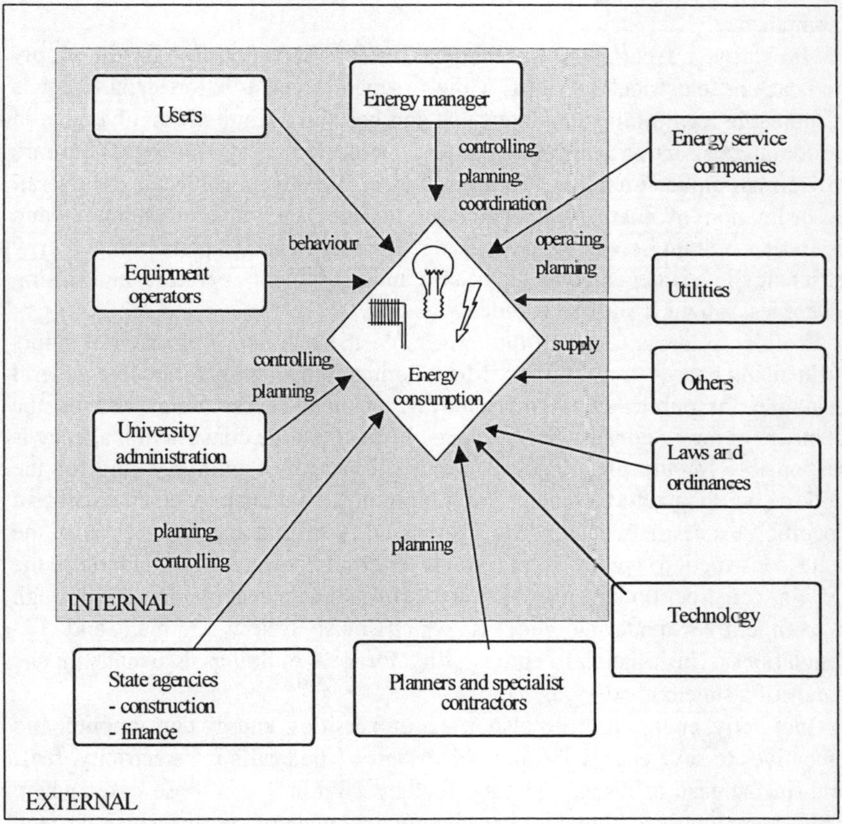

Source: Zimmermann (1998), modified.

Figure 3.1 Determinants of energy consumption at German universities

Within the university, staff and students influence energy consumption directly through switching lights on and off, opening windows, selecting room temperatures, leaving computers running and so on. Similarly, equipment operators exert an influence through controlling the heating, ventilation and air conditioning equipment and setting parameters such as temperature and heating intervals. These may be set manually through local controls, or centrally through the use of electronic Building Energy Management Systems (BEMS). The university administration has a variety of roles, including identifying and prioritising capital investments, measuring and recording energy consumption and costs, purchasing equipment, planning space and (partially) planning and undertaking buildings construction and buildings maintenance.

Typically, a department for technical services is responsible for the supply of heat and electrical services, while a separate construction department is responsible for buildings maintenance and has some (limited) involvement in building construction and refurbishment. Dedicated energy managers are rare in German universities, but if present they will be responsible for the overall co-ordination of energy management activities, including maintaining and operating building controls, identifying, appraising and implementing energy efficiency investments, consulting users and equipment operators, and raising awareness amongst staff and students.

Besides actors internal to the university, there are several external actors influencing energy consumption. Most German universities follow the general principle for public administration which is that users cannot also be the planners of major construction projects. Instead, a state construction agency is responsible for the planning and construction of new buildings and for the bulk of building refurbishment. Even buildings maintenance which exceeds a specific cost level is usually the responsibility of this agency. Normally the state construction agency commissions external contractors to undertake the design, construction and refurbishment work, and selects contractors through a competitive tendering process which must follow German and EU regulations. The financial responsibility for new buildings is usually in the hands of a state budgeting agency.

Indirectly, energy utilities also affect universities' energy consumption and incentives to save energy through the design of the tariffs for electricity, fuels and, in the case of district heating, for heat. Similarly, in those cases where ESCOs are involved, they will have an impact on energy performance through planning and operating energy supply and energy-using equipment through a variety of contractual arrangements. Other external agents include state and federal agencies, ministries and parliaments, who decide upon the financing or co-financing of investments and the establishment of specifications and standards for buildings and equipment. The most relevant federal laws for the HE sector include the Energy Conservation Ordinance, the Heating Installations Ordinance, the Ordinance on Small Combustion Plants and the Ordinance on Thermal Insulation. In addition to the federal laws, there are numerous state laws, decrees and guidelines which affect energy consumption.[7]

The net result is that final energy consumption is determined by the decisions and actions of individuals and institutions at a number of different levels. In each case, decision-making is constrained and shaped by rules, routines and incentive structures which may not operate to the benefit of energy efficiency. Interactions between these groups are complex and there are particular tensions between autonomy and accountability in decision-making.

2.4 Decision-Making Procedure for Investments in Buildings Maintenance

Besides the budget for new buildings and the budget for minor construction work which increases the overall value of the building, the major financial source for energy efficiency is the budget for buildings and estate maintenance. The complex procedure associated with the planning and realisation of projects for buildings and estate maintenance is illustrated in Figure 3.2.

First, users make requests for measures which are collected in a list. Small investments are checked by the university administration – usually the department for technical services – for necessity, urgency and costs. Since, in most cases, the costs associated with the requested measures greatly exceed the maintenance budget, the measures have to be prioritised. The criteria to do so generally include urgency, relevance for teaching and research, health and safety considerations, regulatory compliance and costs. In addition, the head of the university (the President) may request additional measures following new academic appointments. The list of measures is then presented and discussed with the chief administrator of the university (the Chancellor), who is primarily responsible for financial and budgeting affairs. A list then results of measures to be realised. For bigger, more expensive measures or for measures requiring technical and engineering expertise, the state construction agency takes responsibility and commissions designers and contractors to realise the projects. Usually, for these investments, the university administration is no longer involved.

Once a proposal for investment has passed the university administration it goes on to the state administration, where it has to be passed by other ministries (usually the Ministry for Education and Science and the Ministry for Finance). Eventually the state budget has to be approved by the state parliament. New investments for which the federal budget provides co-financing have to be included in the general construction plan for the HE sector (Rahmenplan fuer den Hochschulbau nach dem Hochschulbaufoer-derungsgesetz). The Science Council (Wissenschaftsrat), an academic advisory body at the federal level, gives an expert's opinion of the general plan and makes suggestions. Further approval has to be given by the federal administration and eventually by the federal parliament (upper and lower houses) for the general budgets. Thus the decision-making process for investments in universities tends to be an extremely lengthy, bureaucratic and formal procedure, which consists of various levels not only within an institution, but also within different state and possibly federal ministries and agencies. In addition the exact procedures vary slightly across the fifteen German federal states.

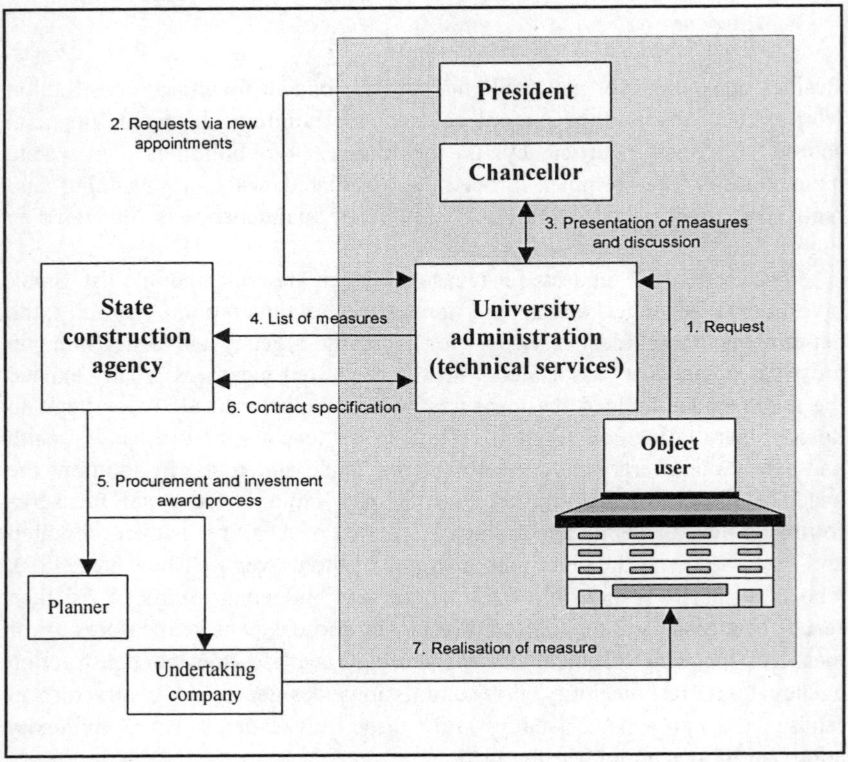

Source: Zimmermann (1998), modified.

Figure 3.2 Decision-making structure for buildings and estate maintenance

3. CASE STUDIES OF ENERGY MANAGEMENT IN GERMAN UNIVERSITIES

Case studies of energy management practices were undertaken in six universities between the spring and autumn of 1999. At these universities, up to six people (top administration, energy managers, technical managers, finance managers, purchasing office, buildings officers and so on) were interviewed in person and, where necessary, follow-up telephone interviews were also conducted. The interviews were semi-structured and used detailed protocols based on the theoretical framework developed in Chapter 2 (see Annex A and B). Table 3.2 summarises the main features of each case study.

The following sections summarise the energy management practices at the case study universities.

3.1 Organisation

Each case study university had a complex structure for decision-making. Only two out of the six universities (A and E) had an energy manager, but this is a higher proportion than in the university population as a whole. Otherwise energy management was one of many responsibilities of the department for technical services. State construction agencies were responsible for construction, refurbishment and, in some cases maintenance of buildings, and typically the influence of the university in these areas was limited. For example, at university A, the state construction agency was responsible for all projects exceeding €10,000 (a very low threshold) as well as for projects where special expertise was required. As a result, in addition to new build and refurbishment, the state construction agency managed more than two-thirds of the funds for building maintenance. Since involvement of the state construction agency entails a lengthy and formal process, it may take years before an approved project is implemented.

Time constraints on staff were a dominant theme in all the case studies. Also, most of the universities exhibited a lack of co-ordination, delegation and clear responsibilities for energy management, even in those institutions that had an energy manager.

3.2 Energy and Environmental Policy

None of the universities had an energy policy and only one (C) had a formal environmental policy. This related to the sustainability concept of Agenda 21 and was in accordance with the CRE-Charter for Sustainable Development (CRE-COPERNICUS). However the policy statement was rather general and did not include either specific targets for environmental

Table 3.2 Summary of German university case studies

Feature	**A**	**B**	**C**	**D**	**E**	**F**
Number of students	16,000	15,000	10,000	11,000	18,000	24,000
Budget in million euros*	200	300	60	70	n/a	250
Net internal area (m^2)	200,000	240,000	76,500	63,800	160,000	530,000
% of pre-1945 buildings	20 %	30 %	70 %	0 %	n/a	33 %
Energy costs in million euros	4.3	5.2	1.1	1.2	n/a	6.0
Energy as % of total budget	2.3 %	1.7 %	1.8 %	1.7 %	n/a	2.4 %
Fuel consumption	82.5 GWh	44 GWh	13.6 GWh	18.2 GWh	n/a	82 GWh
Electricity consumption	36.5 GWh	40 GWh	7.95 GWh	7.9 GWh	n/a	41.6 GWh
Total energy consumption	119 GWh	84 GWh	21.6 GWh	26.1 GWh	n/a	123.6 GWh
Fuel intensity	1.5 GJ/m^2	0.66 GJ/m^2	0.64 GJ/m^2	1.03 GJ/m^2	n/a	0.56 GJ/m^2
Electrical intensity	0.66 GJ/m^2	0.6 GJ/m^2	0.37 GJ/m^2	0.45 GJ/m^2	n/a	0.28 GJ/m^2
Energy intensity	2.16 GJ/m^2	1.26 GJ/m^2	1.01 GJ/m^2	1.48 GJ/m^2		0.84 GJ/m^2
Heat/power ratio	2.3	1.1	1.7	2.3	n/a	2.0
BEMS in place?	not yet	yes, but old	partly	partly	yes	partly
Self-assessment score	3.3	2.0	2.0	1.66	n/a	0.83
% of listed technologies adopted	55%	55%	80%	70%	n/a	40%
Full-time energy manager?	Yes	No	No	No	Yes	No
Environmental policy	No	No	Yes	No	No	No
Payback criteria	5 years, flexible	< 5 years	5 years, flexible	no fixed rules	no fixed rules	no fixed rules

Note: *including third-party funding.

improvement or any structure of responsibility and accountability for realising environmental objectives. University C had, however, developed the basis for an environmental management system and was also the only institution in the sample which aimed to be certified under ISO 14001. The other universities had either not considered this possibility or, as in the case of university A, considered the costs of certification to be much higher than the potential benefits.

As a general – but not formally encoded – rule, teaching and research were the overriding priorities which must not be adversely affected by energy saving measures.

3.3 Energy Information Systems

The extent to which energy information systems were in place and the quality of the installed systems varied considerably across the case study universities. Except for university A, all universities had established BEMS, although usually only for a subset of buildings. At university B, the BEMS was more than twenty years old and largely incompatible with new control technologies. Thermal energy consumption was normally metered monthly at the level of individual buildings or building complexes, but for a small percentage of the buildings (<10 per cent), thermal energy consumption was estimated on the basis of floor area because the cost of sub-metering was considered prohibitive. For some energy-intensive equipment, electricity consumption was individually metered. Energy consumption data was not available at the level of individual departments and while energy consumption was controlled, no targets had been established at either the department or university level. Cost savings from energy efficiency measures were explicitly recorded only at universities A and D.

3.4 Accountability and Incentives

No university directly charged departments or institutes for energy costs. Instead these costs were paid out of a central budget. Plans existed for the introduction of devolved budgeting as part of the new business accounting system, but this will be problematic as sub-metering is incomplete. Allocation based on space occupied may be used as a (rather poor) proxy. At the time of the case studies, therefore, there were no financial incentives for individual departments to save energy.

Importantly, there was a similar lack of incentives to reduce energy consumption for the university as a whole. This was due to the accounting system for public administrations (*Kameralistik*), which requires separate budgets for investments and operating expenditures to be agreed prior to each financial year. In principle, designated but unused funds from one budget (for example, operating expenditure) cannot be used for the other type of budget (for example, investment). Even within the same type of budget, transferability of funds is extremely limited. As a consequence, saved energy costs cannot be easily used to fund further investments in energy efficiency, for the recruitment of personnel or for any other category of expenditure. Similarly, if the university saves energy consumption and thus energy costs, those cost savings will only benefit the state budget, while the university faces the risk of being allocated a smaller budget for energy in subsequent financial periods. The accounting system therefore creates a direct disincentive to improving energy efficiency. Thanks to the principle of 'global budgeting' which is in the process of being introduced in many HE institutions, transferability of funds is likely to become less restricted in the future. With global budgets, there are stronger incentives to save energy costs within the institutions since the cost savings can be retained and used for other purposes. This is an important and badly needed reform.

3.5 Capital Budgeting and Investment

None of the case study universities had a dedicated budget for energy efficiency investment, despite a legal requirement that 6 per cent of the maintenance budget be spent on energy saving measures. In practice, this requirement did not mean much since, first, almost any measure could be declared an energy savings measure; and second, energy efficiency typically improved as a side effect of routine maintenance, owing to technical improvements in energy-using technologies.

The quality of investment appraisal for energy efficiency measures varied considerably across the case study universities. Generally, simple paybacks were used instead of net present value or internal rate of return, and payback periods of 3–5 years were required (Table 3.2). Longer payback periods were sometimes employed for equipment with a long lifetime. In most cases, profitability or risk assessments were of secondary importance for investment decisions. Instead, criteria such as urgency, research requirements, teaching requirements and prestige tended to dominate. Only a few investments were specifically implemented to save energy, and the majority of improvements were obtained through routine maintenance and refurbishment.

As illustrated in Section 2.4, the decision-making process for investments was extremely formal and involved actors not only from within the institution, but also from state and federal ministries and agencies.

3.6 Purchasing and Policy Integration

The most usual arrangement for equipment purchasing was for a central purchasing office to order equipment according to the specifications provided by individual departments. Energy efficiency may be included in the specification, but this was rare. Instead, capital rather than life-cycle cost tended to guide purchasing decisions, partly because departments were accountable for the former but not the latter. The influence of energy managers on purchasing decisions was very limited and the equipment choices made by professors were rarely questioned. Furthermore, the laws and guidelines for integrating environmental performance into purchasing specifications were weak and rarely enforced.

3.7 Awareness and Culture

A general lack of awareness of energy performance and energy costs was prevalent in all the case study universities and awareness raising activities were limited and uncoordinated. Where educational programmes did exist, research and teaching staff tended to abstain because they had 'better things to do'. Motivating students was also difficult because they were only temporarily part of the system and had little incentive to get involved. While reward schemes for students and staff were not seen as an efficient means to generate new ideas, they could play some role in raising awareness. Rewards did not have to be financial for these programmes to work: public acknowledgement and praise was likely to be sufficient.

Support for energy efficiency from the top administration, such as in university A, proved important for getting energy efficiency measures successfully through the decision-making process. However, in terms of generating ideas and identifying potential energy saving measures, the personal motivation and competence of those in charge of energy management was the most important factor. A good example was the head of the technical department at university C, who was also actively involved in the local Agenda 21 process.

3.8 Status of Energy Management

In general, energy management had a very low status compared to research and teaching. However, the case study results suggested a correlation between the status granted to energy management activities and the implementation of energy efficiency measures. The assessed status was based on the subjective judgement of the people interviewed, the extent to which top administration were interested in energy issues and whether there were environmental or energy guidelines. According to this ranking, universities A and C were the only institutions where the status of energy management was rated as being relatively high. At these universities, more weight was attached to energy management when it came to prioritising investment decisions within the given budgets.

4. BARRIERS TO ENERGY EFFICIENCY IN THE GERMAN HIGHER EDUCATION SECTOR

The case study interviews confirmed the initial hypothesis that there are numerous opportunities for cost-effectively improving energy efficiency in the HE sector, but many are not being realised. Most interview partners agreed with the following statement from the pre-interview questionnaire:

> 'There are a wide range of energy efficiency measures that could be implemented in my university that would yield paybacks of less than four years at current energy prices.'

In general, the German case studies suggest that the HE sector manages energy much less effectively than the brewing sector. Performance is comparable to the mechanical engineering sector, but the nature of the barriers are different – notably in the relative importance of split incentive problems and rigid public sector rules.

This section assesses the relative importance of various barriers to energy efficiency in the German HE sector using the theoretical framework developed in Chapter 2. When interpreting the results, it is important to note that the barriers represent perspectives that highlight particular aspects of a complex situation and that there is much overlap and interdependence between the different categories.

The barriers found to be of particular importance are split incentives, lack of capital, hidden costs and imperfect information. These are discussed in turn below. Table 3.3 specifies the particular form in which these barriers operate, while the final column in Table 3.3 identifies potential policy measures to address the respective barrier.

4.1 Split Incentives

Unlike the brewing and mechanical engineering sectors (Chapters 5 and 6), the most important barriers to energy efficiency in the German HE case studies resulted from various forms of split incentives. These could be addressed by properly designed contracts, but this may entail transaction costs.

First, at the university level, there was no incentive to save energy costs since the savings were not allowed to be used for other purposes and may instead lead to reductions in future budgets. At the root of this problem are public accounting principles which limit the transferability of funds. Unused funds designated for a particular purpose can neither be spent on other purposes nor transferred across budgeting periods. For example, savings in energy costs may not be spent on investment in energy efficiency, buildings maintenance or office equipment. Historically, this was the case in all universities. At the time the interviews were conducted (1999), budgeting principles were in the process of being changed in most federal states. However, the policies and future plans on how to allocate savings differed across states. For most universities, partial transferability of unspent funds to other purposes within the same budgeting period was being planned for the near future in connection with the introduction of 'global budgeting'. Similarly, for savings at the university level, some type of cost sharing arrangements with the state administration were being planned, but not all the savings made in the past will remain at the universities for future periods. Thus, incentives for the university to save energy costs are likely to improve, but remain constrained.

Second, energy consumption was not always metered for individual departments and historically there had been no arrangements for decentralised accountability of energy costs. Instead these costs were paid out of the university-wide budget for operational costs. Since energy budgets and the responsibility for energy management were not devolved to individual departments, these had no incentive to save energy through purchasing efficient equipment, office and space management, or behavioural measures. Again, these are typical split incentive problems and they were encountered in all the universities studied. With the introduction of the business accounting system in most universities, it is planned to allocate energy costs in future, most likely in accordance with some proxy such as floor space. While this represents a poor second-best to directly charging for energy consumption, it should nevertheless create stronger incentives than the existing mechanisms to manage office and room space. One interviewee noted that:

> '...so far the use of space is considered free and not associated with any costs,... and the status of a professor is also determined by the office space he manages to seize,...Institute managers will be stunned once they actually have to face costs, especially energy costs.'[8]

Table 3.3 Barriers to energy efficiency in the German HE sector and proposed policies to overcome them

General category	Specific instance	Comments	Policies
Split incentives	Departments not accountable for energy costs	No devolved budgeting; costs and difficulty of introducing sub-metering	Devolved budgeting with new business accounting system
	Limited transferability of funds	Saved energy costs cannot be used for other purposes	Global budgeting (in the process of being implemented in most universities)
	State construction agency responsible for the planning of new buildings and refurbishment	Input of university limited; design, prestige and other factors often more important than life-cycle costs; agency accountable for capital cost, but university responsible for operating costs	Increase universities' planning and financial authority; put operating/facility/space management company in charge; privatisation of all facility/space
	Contractors, etc. for new buildings not accountable for operating costs	Incentive and co-ordination problem; by law many trades involved	Targets for energy performance and operating costs to be included in tender; integrated design processes

Access to capital	Availability of capital to university	Tight federal and state budgets; credits restricted	Energy service contracting
	Allocation of capital within university	Other priorities for capital expenditure; low status of energy management	Conduct profitability analyses; life-cycle costing; make university funding a function of energy performance or energy audits; environmental/energy management schemes; subsidies for energy audits; voluntary/negotiated agreements and other measures to raise awareness at top administration level
Hidden costs	Lack of time, management costs	Hidden costs usually not included in profitability analysis; lack of management time; additional indirect energy services required by equipment often not considered	Energy service contracting; full-time energy manager; targeted information programmes; co-operative procurement; federal energy agency; subsidies for energy audits
	Complex and time-consuming decision-making process	Process involves several levels within university, as well as various state and possibly federal agencies	Change responsibilities and shorten processes through laws; devolve financial and decision-making responsibility to individual institutions; involve ESCOs
Imperfect information	Information on energy use and needs	Poor data on energy consumption at buildings level, end-use split often unknown; no centralised knowledge; information on user needs not co-ordinated	Invest in information systems and BEMS; energy manager to improve co-ordination of energy management; improved communication with top management; energy committee; federal energy agency; best-practice programmes

Third, as already pointed out, for all new construction and to some extent for building maintenance as well, a state construction agency other than the university was responsible for the planning and implementation of the projects. Thus, those planning the projects were not the same as those who had to pay the operating costs. Except for university E, which for historical reasons was also in charge of construction planning, all of the universities complained that the state construction agency did not adequately take into account future energy and other operating costs. Instead, other motives like prestige and design quality appeared to be more important. Quite often, this situation caused friction between those responsible for energy management at the university level and state planners, as indicated by the following statements from interviewees:[9]

> '... those deciding have the power, but they do not have the relevant knowledge and information about the actual needs at the university level.'
>
> '...planners at the state construction agency are utterly incompetent and haven't a clue about what is going on.'
>
> 'Often, one hand does not know what the other hand is doing, or why it is doing what it is doing.'
>
> '...input from the universities is not always appreciated at the state construction agency.'
>
> '...it is not important to save energy or energy costs, but rather to comply with regulations.'

These problems have been exacerbated in the past by external designers and contractors, who are commissioned by the state construction agency and reimbursed according to the Ordinance on Fees for Architects and Engineers (HOAI). Since their fees depend on the financial size of the projects, planners have an incentive to choose larger heating, ventilation or cooling equipment than necessary, which leads to higher energy consumption (see Chapter 7). Similarly, additional efforts on the part of the planners to improve energy efficiency were not incentivised or rewarded until recently, with the result that architects and engineers were not reimbursed for their extra work. In 1994, the German Federal Government modified the HOAI to provide financial incentives for improved energy efficiency or the inclusion of renewable energy sources. Although higher fees can now be charged if the services provided exceed those usually required, the reimbursement may not fully cover the costs.

Even with these additional incentives, split incentive problems remain since architects, engineers, sub-contractors and consultants are accountable for capital cost but not for future operating costs. For them, meeting deadlines and staying within the budget are vital. Lack of co-ordination between different trades tends to aggravate this problem (see Chapter 7).

In contrast, other potential forms of split incentives such as landlord–tenant relationships in leased buildings and short-term incentives created by rapid job rotation were not found to be significant in the HE sector.

4.2 Access to Capital

The lack of capital for investment was a major barrier to improving energy efficiency at the case study universities. Access to capital proved to be a problem at the sectoral level as well as within individual institutions

While business risk and low profitability proved to be the major reasons for lack of capital in the brewing and mechanical engineering sectors (Chapters 5 and 6), the situation was different for universities. Here, the financing for larger investments comes from the state and federal budgets, which are very tight. The situation is seriously aggravated by the fact that, historically, universities have not been allowed to borrow on the capital market. Thus taking up a credit for investments in energy efficiency has not been an option, no matter how profitable the investment may have been. Likewise, in some states third-party financing through an energy services contract was considered illegal (see also Section 5.4). Hence, in universities, the funding situation for investments in general is difficult, which tends to restrict funding to small projects with payback times of under five years. Investments in energy efficiency will at best be delayed, as in the case of university C where the implementation of a highly cost-effective lighting system took 11 years! This delay resulted from the combination of lack of capital in the public sector as a whole, and the complex decision-making structures in the HE sector. For most universities, financial autonomy and flexibility is expected to improve when the budgeting principles and accounting systems are changed.

While investment in energy efficiency suffers from aggregate capital restrictions, the lack of funding for these investments is also a consequence of priority-setting inside and outside the university. The primary criteria are quality of research, quality of teaching, urgency and design prestige. Energy efficiency – and in some cases even profitability – are only of minor importance. Hence, it does not come as a surprise that none of the universities had a specific budget for energy efficiency investment. Instead, small investments had to be financed through unused funds from other areas, in particular through the maintenance budget, where laws require that a certain percentage is spent to improve energy efficiency. But in most cases the amount of money available from this source is both small and insufficient.

4.3 Hidden Costs

Three categories of hidden costs were considered: general overhead costs of energy management; costs specific to a technology investment; and loss of utility associated with an energy efficient technology.

The overhead costs of energy management, including those for staff salaries, energy audits, information systems, sub-metering and so on were usually not included in profitability analyses – if such analyses were conducted at all. Typically, only investment costs, direct personnel costs, maintenance and fuel costs were considered. Incorporating the overhead costs of energy management in profitability assessments would require their magnitude to be calculated. This is no easy task, and leads to additional costs of its own. The prevalence of severe time constraints on university employees was suggestive of the importance of salary costs, in that it may be uneconomic to increase staff resources. However, not only were such calculations not made, but the hiring of new staff was often prohibited by rigid public employment schemes (*Stellenplan*) which fix the number of university staff.

As outlined above, the costs associated with complex and time-consuming decision-making processes constituted a major barrier to all types of investment, including energy efficiency investments. For example, the installation of a CHP-plant would have to be planned 5–7 years in advance – during which time, energy markets and CHP technology may have changed significantly.

By contrast, hidden costs specific to a technology investment, such as disruption, hassle and inconvenience were found to be much less important. Major construction or equipment replacement can always be carried out during the long winter or summer breaks. Likewise, loss of utility associated with energy efficient technologies was not considered to be important. However, in many cases the costs for additional energy services were not considered when buying new equipment. For example, at university A, a new transmission microscope was estimated to increase annual operating costs by €1,600. However, the microscope significantly added to cooling and ventilation loads, and when the fixed, operating and maintenance costs for the cooling and ventilation equipment were taken into account, the additional costs exceeded €6,000/year. Hence, when indirect costs are neglected, as is typically the case, total costs may be seriously underestimated, with the result that inefficient investment decisions are made.

In summary, the hidden costs associated with identifying savings potentials, finding appropriate energy efficiency measures, conducting profitability analyses and preparing public procurement processes were important barriers to energy efficiency in the case study institutions. But the evidence does not support the hypothesis that these costs are significantly higher for energy efficient measures than for standard measures.

4.4 Imperfect Information

Two categories of information were considered: the volume and pattern of energy use in the universities; and the opportunities available for dedicated energy efficient investments.

Information on current energy use at the level of individual buildings or departments was poor and the end-use split of energy consumption was generally unknown, particularly for electricity. Data on energy costs, energy consumption and user needs was collected by individual departments rather than a central organisational unit and there was no evidence of comparison with either generic or sector-specific benchmarks. Similarly, the needs of end-users were either not known or not well communicated within the institution. But while the case study results suggest that problems of decentralised information, lack of co-ordination and lack of knowledge of user needs are prevalent throughout the HE sector, they also suggest that these are less pronounced in universities with a full-time energy manager.

While lack of information on the volume and pattern of energy use was considered to be an important barrier to energy efficiency, the quality and quantity of information available on energy-specific investment opportunities was generally judged to be good. In the case studies, information on investment opportunities appeared to depend on the competence and motivation of the staff in charge of energy management. With regard to sources of information on energy efficiency measures, the media used included the internet, informal networks, meetings of energy managers, the information system for the higher-education sector (HIS-Higher Education Information Systems GmbH, Hannover), special seminars and workshops, and ESCOs. If used at all, informal networks were considered to be excellent and trusted sources of information. In contrast, there was considerable suspicion of ESCOs.

5. POLICY IMPLICATIONS

The empirical research demonstrates that there is a wide range of barriers to energy efficiency in the German HE sector. This suggests that, if energy efficiency is to be significantly improved, a similarly wide range of initiatives at the organisational, sectoral and national level is likely to be required. A general observation from the case studies is that successful energy efficiency policy has to encompass more than just fiscal measures. It must also take into account the communicative, organisational and co-operative challenge that energy efficiency creates for individual institutions (Ostertag et al., 1998).

In addition to public policy measures, energy service contracts may provide an effective route for overcoming barriers such as lack of capital, time, staff and expertise. Likewise, the involvement of ESCOs may avoid the time-consuming and complex decision-making structures within the HE sector (Viebahn, 1995; 1996). Hence, this section will discuss two broad approaches to overcoming the barriers identified in the previous section. First, some potential measures at the organisational, sector and national level will be proposed. Then, the limiting and success factors for ESCOs in the German HE sector will be identified and assessed.

5.1 Organisational Level

To reduce information deficits, improve the co-ordination of energy management and user needs, and encourage vertical and horizontal communication, the position of an *energy manager* should be established. Along with other benefits, this individual could assume the role of 'product champion' for energy efficiency, which has proven to be instrumental for improving energy efficiency in the industrial sector (Ramesohl, 2000). Likewise, *bundling information and competence for energy use and consumption* within a single department saves co-ordination costs and helps to identify and solve internal conflicts of interest. This means that responsibility for planning, maintenance, technical services, space management, heat and electricity supply and buildings energy management systems should be under the same roof. In any case, *clear delegation and responsibility for energy management is crucial*.

Certified environmental management schemes (EMS) may also be an effective measure, especially for poor performers. These schemes not only organise environmental management, but are designed to motivate staff and students and to get energy and environmental issues onto the top administration's agenda. For large universities, implementing an EMS at the level of the institution requires considerable effort in co-ordination and hence may be difficult to achieve. Environmental management schemes for individual schools or institutes should therefore be considered as an alternative. Getting the top administration's attention may also be achieved via *voluntary or negotiated agreements*, either at the level of the institution or for the entire HE sector. Given the crucial role of the state administrations, voluntary agreements may be most effective for organisations at the state level. Instead of using payback rates, *profitability analyses* using life-cycle costs should be carried out, so the cost-effectiveness of energy efficiency measures can be demonstrated. With modern software tools, the transaction costs of this should not be prohibitive. Similarly, when deciding on new equipment, the *indirect costs for additional energy services* should be considered. Since indirect costs can sometimes be a multiple of the direct investment costs, neglecting them may result in inefficient investment decisions.

5.2 Sectoral Level

Creating positive incentives for universities to invest in energy efficiency implies that institutions should be able to keep the cost savings from these investments and use them for other purposes now or later. That is, universities should be able to shift funds between budget headings and they should not be punished for lower energy costs by receiving less funding overall in subsequent financial periods. Thus *global budgeting with unrestricted transferability of funds* within and across budgeting periods should be introduced in all organisations. To make incentives for the planning and operation of buildings compatible, individual universities should be *given more planning (and financial) authority* at the expense of the state planning agencies. Alternatively, the operation and management of facilities and estates could be *outsourced* to private companies.

If a portion of university *funding* were to be made a *function of prior performance* such as achieved energy savings or of some kind of benchmarking, the incentive to realise energy efficiency measures within the university would be much stronger. This kind of funding system would also force energy costs onto the top administration's agenda. Alternatively, a proportion of state funding could be dependent upon the universities having carried out an energy audit within the last year or two or on having an environmental/energy management system in place.

Similarly, individual departments should have *individual budgets* and be held accountable for their energy costs as far as possible. Clearly, allocation of costs based on the space used would only be a first step. At the same time the *business accounting system*, which is currently being introduced in many universities, is expected to lead to a better allocation of financial resources and to provide better incentives for energy savings measures than the obsolete *Kameralistic* system. The switch in accounting systems also provides opportunities to incorporate data on environmental performance into the new system.

Future operating costs should be an integral part of the *procurement specifications* for new buildings and major refurbishment. Likewise, *integrated planning* should be implemented as a rule. Since integrated planning involves all the relevant actors including architects and specialist engineers, sub-contractors and the energy manager, the content of the individual work-packages can be better co-ordinated and designed such that split incentives are accounted for and a more efficient solution emerges (see Chapter 7).

Other possibilities at the sectoral level include *training seminars* for those in charge of energy management as well as *targeted information programmes* on specific topics (for example, contract energy management). Both are expected to reduce information and other transaction costs. In the same way, informal networks such as eco-campus net[10] can be a cost-effective tool to reduce information-related barriers.

Benchmarking at the sectoral level is an important tool to get the top administration's attention and may help improve the status of energy management. However for meaningful comparisons across different institutions, proper benchmarking would have to differentiate between a wide range of buildings types and uses.

Finally, *co-operative procurement* of energy efficient equipment by several HE institutions provides another route to reduce transaction costs and obtain price reductions on equipment through bulk purchase. Recently, primary and secondary schools in the city of Hamburg have successfully used such a procurement process for more energy efficient lighting.

5.3 National Level

The effectiveness of the measures at the organisational and at the sectoral level will be enhanced if they are embedded in a broad-based and long-term national climate programme. In particular, allowing energy prices to reflect external environmental costs will render measures improving energy efficiency more cost-effective and raise awareness among higher administration. In Germany, such policies include the continuation of the *ecological tax reform*, which was implemented in 1998. Increasing tax rates on fuels and electricity reduces the financial risk associated with investments in energy efficiency and allows for long-term planning.

Increasing energy costs will also ensue from *emissions trading schemes* for greenhouse gases. The framework for this was created by the 1997 Kyoto Protocol (Skea, 1999), and starting in 2005, participation in an EU-wide CO_2 trading system (EU ETS) will be mandatory for large installations in the energy industry and carbon-intensive manufacturing industries (European Commission, 2003). After 2008, a wider range of sectors and gases is expected to be covered. A small number of universities will be included within Phase 1 of the EU scheme (2005–2007) as a result of having combustion plant with an aggregate input capacity exceeding 20MW, while others will face higher heat prices as a result of district heating plants being included in the scheme. All universities will face higher electricity prices, since the electricity generators are expected to pass on the opportunity cost of allowances to electricity consumers (Sorrell, 2002). These price increases could be significant and should create additional incentives for energy efficiency.

National policies in *support of CHP* may affect universities either directly, as operators of CHP plants, or indirectly as end-users who bear the costs of national subsidy programmes. After the liberalisation of the German electricity market in 1998, CHP was seriously threatened by the decline in electricity prices. To provide temporary support and to avoid stranded investments, the German government passed the so-called 'KWK-Vorschaltgesetz' (Cogeneration Act) in Spring 2000. This required mandatory feed-in tariffs for electricity generated by municipally-owned CHP plants, but not for industry or state-owned plants. Hence, under the provisions of this law, existing or planned university-owned CHP plants could not benefit. Since then, further regulations have been passed to achieve the target of doubling electricity generation from CHP by 2010. In particular, a voluntary agreement between the government and industry requires industry to invest in CHP while obtaining a premium for exported electricity. If industry fails to meet its commitment, the government plans to introduce a mandatory CHP quota system (BMU, 2001). However most CHP plants at HE institutions do not export electricity. The support scheme would therefore need to be modified if existing CHP plants in the sector were to benefit, or if incentives for new installations were to be created.

Other policies at the national level include the continued *tightening of the standards* in the Thermal Insulation Ordinance or the planned Energy Conservation Ordinance, the re-examination of technical standards for heating, ventilation, air conditioning and cooling services to avoid over-sizing and the introduction of *minimum efficiency standards* and *labels* for energy-consuming equipment such as PCs. Moreover, public programmes *subsidising energy audits* and the implementation of *energy management systems* could be extended to make public institutions eligible. Finally, the recently formed *Federal Energy Agency* could initiate, co-ordinate or develop information and education programmes targeted at the HE sector, together with best-practice programmes, pilot projects, support networks (such as eco-campus net), and 'energy-cocktails' for top administrators, similar to the Swiss RAVEL programme (Bush, 1996).

5.4 Limiting and Success Factors for ESCOs in the German Higher Education Sector[11]

As discussed earlier and illustrated in Box 3.1, energy service contracts may be effective in overcoming key barriers to energy efficiency in the HE sector, such as lack of capital or various types of transaction costs. Following the liberalisation of the German electricity market in 1998, a large number of new energy service companies (ESCOs) have emerged. In 1998, the total market for energy services in Germany amounted to €1.8 billion (Muehlstein, 2000). By far the largest business areas are heat services and CHP-projects, which account for about 80 per cent of total investment. Demand-reducing technologies, such as efficient lighting, insulation or controls or other supply technologies, such as compressed air and process heat, account for only a small fraction.

In the following subsections, limiting and success factors for ESCOs in the German university sector are identified. The findings are based on the case studies and on additional interviews with selected ESCOs.

Limiting factors

One of the most important reasons why ESCOs are reluctant to offer contract energy management in the HE sector is the long negotiation time before a contract can be completed. The decision-making process has to follow administrative procedures and budgeting laws, which are complex, inflexible and time-consuming. In addition, when considering whether to use an energy services contract or whether to realise the project internally and keep the profits, universities tend to underestimate the hidden costs associated with realising the project under their own authority. Similarly, some universities compare new offers from ESCOs with the variable costs of providing the same services internally. This incorrect procedure renders offers by ESCOs, which are calculated on the basis of full costs, seemingly unprofitable. Furthermore, outsourcing is often associated with the loss of jobs in the public sector, which creates a political problem for decision-makers. Even if university employees are taken over by the ESCO, problems may arise from the change in the nature of the employees' contracts from public to private law. Employees are reluctant to give up job security or other benefits provided under public sector employment.

Box 3.1 The energy services concept

Energy services represents a new and rapidly growing business model in which suppliers offer a single contract to minimise the total bill for the services that energy provides, such as heating, lighting and air conditioning. This contrasts with the traditional model in which energy consumers contract separately for energy commodities and for a range of energy conversion equipment. In its simplest form, an energy services contract may guarantee supplies of heat and power at reduced cost, but in a more sophisticated form the contract may guarantee particular levels of service provision (for example, lighting levels, room temperatures).

By focusing on better performance and solutions to customer needs rather than commodity sales, energy service companies (ESCOs) have a strong incentive to improve energy efficiency. ESCOs typically offer energy management, energy information systems, energy audits, installation, operation and maintenance of equipment, competitive finance, and fuel and electricity purchasing. The contract allows the host organisation to reduce risk, avoid capital expenditure, reduce energy costs and concentrate attention on core activities. Energy services could provide a cost-effective route to overcoming barriers to the diffusion of both established and innovative low carbon technologies in the public, commercial, industrial and household sectors. The model is applicable to both energy use and energy supply technologies and is, in may countries, the primary mechanism for the diffusion of CHP technology. It is also consistent with energy market liberalisation and the broader trend towards the outsourcing of non-core activities. As climate policy develops, energy services have the potential to develop into broader *carbon services*, which may include carbon offsetting, renewable energy purchasing, fuel switching and participation in emissions trading.

Important barriers for ESCOs in the HE sector also arise from the public accounting system. Historically, universities have had difficulties in getting the investment component of an energy services contract 'reimbursed' by the state administration. Likewise, some state administrations have denied approval of contract energy management because it was regarded as a type of unauthorised 'hidden credit'. Finally, barriers also include lack of information about contract energy management, lack of credibility and trust with regard to ESCOs, the fear that savings potentials identified by ESCOs will reflect poorly on those in charge of energy management at the university, and the belief that university staff have a better knowledge of user needs. Universities may also be reluctant to engage in long-term contracts because of the increased uncertainty following energy market liberalisation.

Success factors

Since HE institutions belong to the public sector, they carry a very low financial risk. This makes the HE sector attractive for ESCOs, since they typically require long contract periods. As mentioned above, universities have just begun to enjoy increased autonomy in decision-making and financial resource allocation. These developments are expected to speed up decision-making processes and reduce uncertainty stemming from hierarchical administrative structures. This process is being accompanied by a change in the accounting system. At present the public accounting system does not identify expenses and revenues – as the business accounting system does – but rather expenditures and receipts. Thus actual costs in terms of resource use are not captured. The introduction of the business accounting system is expected to facilitate appropriate appraisals of alternative investment projects, which should also help to correctly assess the costs of contract energy management vis-à-vis other alternatives.

Universities are also considered attractive partners for energy service contracts, since their energy consumption is sufficiently large for ESCOs to recover overhead and transaction costs and make a profit. In addition, the homogeneity of energy supply and demand technologies facilitates business strategies focusing on particular customer groups and allows economies of scale and scope to be realised. Finally, support for ESCOs also arises from new 'lean' management strategies at the university level and from state administrations which are beginning to push contract energy management as a strategy to overcome financial restrictions.

6. SUMMARY AND CONCLUSIONS

The German HE sector has significant untapped potential for cost effective energy efficiency improvement. Those in charge of energy management at universities are aware of this potential, but its realisation is prevented by a range of barriers.

Various forms of split incentives were found to be the most relevant barrier to energy efficiency in German universities. First, individual departments or institutes are not accountable for energy costs. Second, budgeting laws prevent cost savings from energy efficiency investment from being spent on other purposes. Third, the fact that state construction agencies rather than individual universities are in charge of new construction and refurbishment means that user needs, university objectives and lifetime costs are not properly taken into account. And fourth, since contractors are not accountable for the future operating costs of a building, they have only limited incentives to improve energy efficiency.

The second biggest obstacle to energy efficiency in the HE sector is the transaction costs associated with complex, hierarchical and time-consuming decision-making procedures. This process not only involves several levels within individual institutions, but also diverse state and possibly federal agencies and ministries.

The case studies also suggest that universities suffer from a lack of capital due to a combination of tight federal and state budgets and legal restrictions on accessing the capital market. Other relevant barriers include inadequate staff resources for energy management (leading to time constraints), poor information on energy consumption patterns due to inadequate investment in information systems, and the low status and insufficient coordination of energy management.

Effectively overcoming these barriers is likely to require a wide range of organisational, sectoral and national initiatives. Clearly, national initiatives such as energy pricing can encourage and support organisational changes. While many of the required initiatives are well established within German energy policy (Schleich et al., 2001b), they must go hand in hand with broader changes in university funding, decision-making and accounting procedures. These 'global budgeting' reforms should lead to increased financial flexibility and autonomy for institutions, a shift of competencies from the state to the institutional level, increased transferability of funds, and the introduction of business accounting. In addition, ESCOs are well positioned to address barriers such as transaction costs, lack of capital, bounded rationality and information deficits, and reforms to university funding should encourage the expansion of ESCO contracts in the HE sector.

Overcoming barriers to energy efficiency and improving energy performance in the German HE sector is likely to be cost-effective, and will also contribute to achieving international and national climate policy targets. Under the EU Burden Sharing Agreement, Germany has committed itself to a 21 per cent reduction in greenhouse gas emissions from 1990 levels. This is by far the largest absolute reduction requirement of any EU Member State and accounts for roughly three-quarters of the required reductions for all member states. By 2000, Germany had already reduced its greenhouse gas emissions by over 18 per cent and is one of only two Member States that are on track to meet their Kyoto targets (the other is the UK). The national greenhouse gas target is even more ambitious: in the early 1990s, Germany committed itself to reduce its CO_2 emissions by 25 per cent by 2005 compared to 1990 levels. Unlike the Kyoto target, this target is likely to be missed unless additional measures are implemented in the near future. The analyses presented in this chapter suggest that the HE sector could contribute significantly and at low cost.

NOTES

1. Case study interview, Spring 1999.
2. As a result of ~15 per cent excess capacity in electricity generation, power companies engaged in fierce price competition on the basis of short to medium-term marginal costs (*Financial Times Energy Economist*, 2000).
3. The remainder broke down as follows: 7 polytechnics, 16 theological colleges, 6 educational colleges, 46 art academies, 152 advanced technical colleges and 36 administration colleges.
4. A range of sources was used to estimate energy consumption and costs, including: Energieverwertungsagentur (1999); Prognos (1999); Umweltbundesamt (1999); and Statistisches Landesamt Baden-Wuerttemberg (1997).
5. This number is lower than recent estimates would suggest (Winkelmann, 2000; Volkens and Schomaker, 2000). In the energy balances, the higher education sector is part of the education sector, which in turn belongs to the aggregate sector 'trade, commerce and services'. In 1997 this sector accounted for 1540 PJ of final energy consumption, which corresponds to about 16 per cent of total final energy consumption in Germany. The entire education sector accounts for about 7.2 per cent of final energy consumption in this sector and thus for about 1.2 per cent of total final energy consumption in Germany (Prognos, 1999). Hence, these figures support our estimates for higher education which imply that this sector is responsible for about one-third of final energy consumption in the education sector.
6. For specific examples of measures realised to improve energy efficiency in universities see Starnick and Winzer (1994), Person (1999), Viebahn/Matthies (1999) or eco-campus net (http://www.eco-campus.net).
7. For an overview for each state see Person (1999).
8. Case study interview, summer 1999.
9. Case study interviews, spring and summer 1999.
10. See http://www.eco-campus.net.
11. This section draws heavily on Schleich et al. (2001a).

REFERENCES

BMU Bundesministerium für Umwelt, Naturschutz und Reaktorsicherheit (2001), 'KWK-Foerderung das richtige Signal vor der Klimakonferenz in Bonn', press release (132/01), Berlin.

Bush, E. (1996), 'Energie-Aperos – Energiesparmöglichkeiten lustvoll vermitteln', *Impuls,* 19, Zeitschrift for IP BAU, RAVEL und PACER, Bundesamt für Konjunkturfragen, Bern.

Energieverwertungsagentur (1999), 'Energieeffiziente Universitaeten', Endbericht, Vienna, April.

European Commission (2003), 'Directive 2003/87/EC establishing a scheme for greenhouse gas emissions allows trading within the Community and amending Council Directive 96/61/EC', *Official Journal of the European Communities*, L. 275/32-46, Brussels.

Financial Times Energy Economist (2000), 'German prices at cash cost', No. 223.

Muehlstein, J., 'Was koennen Contractoren der Industrie bieten', Vortragmanuskript, Forum Hemmnisabbau 27./28.3.2000, Fraunhofer Institute for Systems and Innovation Research, Karlsruhe.

Ostertag, K., U. Böde, E. Gruber and P. Radgen (1998), 'Erfolgreiche Beispiele für die Überwindung von Hemmnissen der rationellen Energieanwendung in Industrie und Kleinverbrauch', Fraunhofer Institute for Systems and Innovation Research, Karlsruhe.

Person, R.D. (1999), 'Rationelle Energieverwendung in Hochschulen', *HIS Hochschulplanung*, 139, HIS GmbH, Hannover.

Planungsausschuss fuer den Hochschulbau (2000), 'Rahmenplan fuer den Hochschulbau nach dem Hochschulbaufoerderungsgesetz 2000–2003', Allgemeiner Teil und Anhang.

Prognos Institut (1999), 'Die laengerfristige Entwicklung der Energiemaerkte im Zeichen von Wettbewerb und Umwelt', Basel.

Ramesohl, S. (2000), 'Social interactions and conditions for change in energy-related decision-making in SMCs – an empirical socio-economic analysis', paper presented at the IPCC Workshop on Conceptual Frameworks for Mitigation Assessment from the Perspective of Social Science, 21–22 February 2000, Karlsruhe.

Schleich J. and U. Boede (2000), 'Reducing barriers to energy efficiency in the German third level education sector', report to DG Research on the project 'Barriers to Energy Efficiency in Private and Public Organisations', Fraunhofer Institute for Systems and Innovation Research, Karlsruhe.

Schleich, J., U. Boede, D. Koewener and Radgen, P. (2001a), 'Chances and barriers for energy services companies: a comparative analysis for the German brewery and university sectors', Proceedings of the 2001 ECEEE summer study, 11–16 June 2001, Mandelieu, France.

Schleich, J., R. Betz, F. Gagelmann, E. Jochem and D. Koewener (2001b), 'Germany: unification and contradiction', in C. Vrolijk (ed.), *Climate Change and Power. Economic Instruments for European Electricity*, Royal Institute of International Affairs, Earthscan, London.

Skea, J. (1999), 'Flexibility, emissions trading and the Kyoto Protocol', in S. Sorrell and J. Skea (eds), *Pollution for Sale: Emissions Trading and Joint Implementation*, Edward Elgar, Northampton, MA and Cheltenham, UK.

Sorrell, S. (2002), 'The climate confusion: implications of the EU emissions trading scheme for the UK Climate Change Levy and Climate Change Agreements', SPRU – Science and Technology Policy Research, University of Sussex, Brighton, available from: http://www.sussex.ac.uk/spru/environment/research/ interact.html.

Sorrell, S., J. Schleich, S. Scott, E. O'Malley, F. Trace, U. Boede, K. Ostertag, and P. Radgen (2000), 'Barriers to energy efficiency in public and private organisations', final report to DG Research under the project 'Barriers to Energy Efficiency in Public and Private Organisations', SPRU – Science and Technology Policy Research, University Of Sussex, Brighton.

Starnick, J. and M. Winzer (1994), 'Umweltschutz an deutschen Hochschulen', *Handbuch fuer Umweltschutz*, Hochschulrektorenkonferenz (ed.), Alpha-Informationsgesellschaft mbH, Bonn.

Statistisches Bundesamt (1999a), *Statistisches Jahrbuch*, Wiesbaden.

Statistisches Bundesamt (1999b), *Hochschulstandort Deutschland*, Wiesbaden.

Statistisches Landesamt Baden-Wuerttemberg (1997), 'Das Bildungswesen 1997', *Statistik von Baden-Wuerttemberg*, Band 526, Stuttgart.

Umweltbundesamt (1999), 'Umweltcontrolling im Bereich der oeffentlichen Hand', UBA Texte No. 8/99, Berlin.

Viebahn, P. (1995), Energieeinsparung durch Nutzlight-Contracting. Ein Einsparkraftwerk für die Universitat Osnabrück und die Widerstände dagegen. Paderborn: Möllmann.

Viebahn, P. (1996), 'Energieeinsparung durch Nutzlight-Contracting. Entwicklung eines contracting-Modells für das Allgemeine Verfügungszentrum der Universität Osnabrück', Wuppertal-Papers Nr. 60, Wuppertal Institute.

Viebahn, P. and M. Matthies (1999), *Umweltmanagement an Hochschulen: Konzepte, Strategien, Loesungen*, projekt verlag, Bochum.

Volkens, A. and K. Schomaker (2000), 'Der Beitrag der Hochschulen zu einer nachhaltigen Entwicklung', *Oekologisches Wirtschaften*, 3-4, 1–2.

Winkelmann, H.-P. (2000), 'Energieeinsparung an Univesitaeten und Hochschulen – Handlungsansaetze fuer eine nachhaltige Entwicklung der Hochschulen am Beispiel der rationellen Energienutzung. CRE COPERNICUS', Dortmund.

Zimmermann, N. (1998), 'Effekte der Effizienzsteigerung in der Energiebewirtschaftung durch die Einfuehrung von Energiebeauftragten in der oeffentlichen Verwaltung am Beispiel der TU Darmstadt', Master's thesis, Economics Department, University of Darmstadt.

4. Leaving Money on the Floor: Barriers to Energy Efficiency in the UK Higher Education Sector

Steve Sorrell

'If I looked at all the opportunities with four year paybacks I would need another four staff…'[1]

1. INTRODUCTION

The UK higher education (HE) sector provides an extremely rich example of barriers to energy efficiency. This sector is interesting for its diversity of energy-using technologies, for the size and scope of energy efficiency opportunities, and for the complexity of decision-making processes within large public sector organisations, with all the attendant problems of capital constraints, split incentives and conflicting objectives.

With energy use dominated by generic applications such as heating and lighting, and with energy costs accounting for only ~1.7 per cent of total expenditure, HE can be taken as representative of much of the public and commercial sector. Energy is given a low priority by management and the great majority of staff are neither accountable for energy costs nor concerned about energy issues. The constraints imposed by public sector funding are clearly evident, as is the impact of reform initiatives to improve efficiency and accountability and to mobilise private sector finance. The sector is also a major client for the construction of new buildings and has an extensive programme of refurbishing existing buildings. It is these capital projects that provide the greatest opportunity for improving energy efficiency.

This chapter explores the extent to which UK universities are neglecting energy efficiency opportunities and the validity of competing explanations for this. The study is based upon a postal survey of energy management in the sector, together with six in-depth case studies of individual universities. The chapter begins with an overview of the sector, including the financial and organisational context, the pattern of energy use and the recent initiatives on energy and environmental management. The results of the postal survey are then reported, followed by a description and assessment of energy management practices in the case study institutions. These results are then combined to identify those barriers which appear of particular importance in the sector and to assess how they operate. This leads to suggestions on how the barriers may be overcome and how energy efficiency in the sector may be improved.

2. THE UK HIGHER EDUCATION SECTOR

There are 90 universities in the UK, or 114 if the colleges of the federal universities of London and Wales are counted separately. There are also 35 smaller HE colleges and 45 specialist institutions, but the focus in this chapter will be the universities. In 1998–99 there were 1.2 million full-time HE students and 0.6 million part-time students taught by 0.13 million academic staff (HESA, 2000). Some 25 per cent of full-time students were housed in university accommodation.

Total expenditure by the HE sector in 1998–99 was £11.9 billion[2] out of an income of £12.1 billion (Ramsden, 2001). This gave an operating surplus of 1.6 per cent of income, close to the average over the previous five years (1.8 per cent) but substantially less than the 3 per cent recommended by the government funding bodies to fund reinvestment and future developments (HEFCE, 2000a). In addition to restricting the scope for funding capital expenditure through retained earnings, this leaves universities vulnerable to small changes in forecast income or expenditure. Financial strength is also very unevenly distributed, with 10 institutions holding 50 per cent of the sector's cash reserves, and with 30 institutions in deficit (HEFCE, 2000a).

The primary (41 per cent) source of funding is the Higher Education Funding Councils (HEFCs), which distribute public funds for teaching and research according to quality criteria.[3] Additional funding is provided by the Research Councils, which fund the direct costs of individual research projects. Public funding of infrastructure has been progressively cut back over the last 20 years, and since 1994 the HEFCs have ceased to allocate capital grants for new buildings except for certain exceptional circumstances. As a result, universities now fund capital projects through a variety of sources, including sponsorship and donation (NAO, 1998). Capital expenditure in 1998/99 was £1.02 billion (8.4 per cent of income), of which 22 per cent was through capital grants, 39 per cent through internal sources and 31 per cent through new borrowing (HEFCE, 2000a).

Over the last 30 years, UK higher education has evolved from an elite system, with 5 per cent of young people attending university in 1960, to a mass system where one in three school leavers now study for a degree. This has created enormous pressure on resources as universities attempt to maintain traditional degree structures and teaching styles. The combination of constraints on public expenditure and rapid growth in student numbers led to a reduction in unit funding per student of 40 per cent between 1977 and 1997. Unit cost reductions were partly achieved by under investment in infrastructure, leading to a funding gap for capital projects and an increasing crisis. Following a Committee of Inquiry in 1997 (Dearing, 1997), the government introduced means-tested tuition fees, allocated an additional £450 million to the sector and promised full funding for the expected increase in student numbers over the following three years (HMG, 1998). These initiatives have ameliorated the difficulties to some extent, but the sector still faces serious problems with backlog maintenance, estimated at £3.4 billion in 2000 (HEFCE, 2001).[4]

2.1 Energy Use in Higher Education

The HE sector presents both a physical and organisational challenge to energy management. Physically, many universities consume as much energy as a small town, with bills exceeding £3million/year. Consumption is dominated by heating, ventilation, air conditioning (HVAC) and lighting, but these are distributed between an enormous range of building types, uses and ages. Most universities will have offices, catering facilities, sports facilities, residences and laboratories, while many will have specialist scientific equipment with high energy loads. Two relevant distinctions are:

- between 'old' (pre-1992) universities, many of which have 'listed'[5] buildings, and 'new' universities (previously polytechnics) which were formerly under local authority control;[6] and
- between campus universities, located on large out-of-town sites and often with district heating systems, and metropolitan universities comprising a larger number of smaller buildings distributed throughout a metropolitan area.

UK universities have complex decision-making structures, built around a series of committees, with ultimate authority lying with the University Council or equivalent. Organisation and procedures may differ between old and new universities, as may the relative priority given to teaching and research. In all cases, university procedures must comply with guidelines laid down by central government and the HEFCs, including the Financial

Memorandum (HEFCE, 2000b). Energy and water management are normally brought together under the heading of 'utilities', centralised in the Estates Department ('Estates') or equivalent, and may or may not involve dedicated energy management staff. Estates are also responsible for new construction, refurbishment, routine maintenance and energy purchasing.

Table 4.1 provides some useful data on the composition, quality and costs of the UK university estate, including energy costs. Total expenditure on energy in the sector is approximately £200 million/year, which represents around 1.7 per cent of total expenditure by the sector, or 4 per cent of government spending on HE.

Table 4.1 Estates in UK higher education – 2000–2001

Total revenue property expenditure	£1238 million
Maintenance expenditure	£412 million
Capital expenditure	£792 million
Total backlog cost of maintenance	£3444 million
% of non-residential estate requiring major repair	37%
Total property costs	£81.05/m^2
Capital expenditure	£29.77/m^2
Maintenance costs	£16.78/m^2
Estate management costs	£2.79/m^2
Energy costs	**£6.10/m^2**

Note: Unit figures refer to net internal area.

Source: HEFCE (2001).

Information on the pattern of energy use in HE is relatively poor, but aggregate data is available on energy consumption by end-use for the education sector as a whole (Bruhns et al., 1997; Mortimer et al., 1998). This shows that fossil fuels accounted for 85 per cent of energy use in the education sector in 1996, of which 74 per cent was for space heating. Lighting was the biggest electricity consumer, accounting for 61 per cent of the total, while air conditioning accounted for 1 per cent and IT equipment only 2.3 per cent. This compares to 11 per cent and 12 per cent respectively in commercial offices. Energy use in HE may differ substantially from that in the education sector overall, since universities may be expected to have longer hours of occupancy, a greater diversity of end-uses, higher specific consumption, and increasing electricity demand for air conditioning and IT.

The average energy intensity of UK universities is estimated to be 1.1 GJ/m^2 (+0.2/-0.1GJ), which is high compared to other non-domestic buildings

(Ashley et al., 1997). For example, universities are 60 per cent more energy intensive than commercial offices ($0.68GJ/m^2$), and more than twice as energy intensive as manufacturing premises ($0.43GJ/m^2$). More disaggregated data is available from the HEFCs, which have identified benchmarks for energy intensity for a range of building types and activities (Table 4.2) (HEFCE et al., 1996c).[7] These show that:

- There is very wide variation between building types and activities. For example, a good consumption level for a naturally ventilated administration building is around 0.21 GJ/m^2 while the corresponding figure for fast food catering is $1.58GJ/m^2$.
- Air conditioned buildings typically consume more than twice the amount of energy as naturally ventilated buildings, most of which is electricity.
- Science laboratories use around 60 per cent more energy than arts academic areas.
- Residential areas use around twice as much fuel as academic areas, but less electricity.
- Electricity dominates energy costs. For example, an air conditioned library can have five times the energy costs of a naturally ventilated library.

2.2 Energy Saving Potential

There have been no detailed studies of the energy saving potential within the UK HE sector. A 1996 study by the funding councils suggested that savings of 10 per cent to 20 per cent were possible 'over a period', but did not indicate either the nature of the specific opportunities or the investment criteria assumed (HEFCE et al., 1996a). However, there are a range of studies on the energy efficiency potential within the service sector overall (for example, Mortimer et al., 1998; IPCC, 2001). Table 4.3 is taken from authoritative UK data (Fisher et al., 1998) and distinguishes between the technical potential resulting from the retrofit of 25 specific, off-the-shelf energy saving techniques,[8] and the economic potential, calculated using an 8 per cent rate of return and ignoring hidden costs. This suggests that a quarter of sector energy consumption could be saved through investment with an 8 per cent rate of return.

As with other studies of this type, a range of caveats apply. While the neglect of hidden costs and other barriers could bias the estimate of economic potential upwards, it is equally possible that the neglect of technical change and the restricted level of technology disaggregation could bias the estimate downwards. Some authors argue that such energy models use an incremental approach which neglects the potential savings from the

Table 4.2 HEFCE benchmarks for energy consumption in higher education buildings

Designated Activity Area	Subtype	Electricity GJ/m^2		Fossil Fuel GJ/m^2		Total GJ/m^2	
		low	high	low	high	low	high
Catering	Fast Food	0.72	0.78	1.58	2.22	2.30	3.01
	Bar/Restaurant	0.49	0.54	0.66	0.93	1.15	1.46
Academic	Science – Laboratory	0.56	0.63	0.40	0.48	0.95	1.11
	Science – Other	0.41	0.46	0.40	0.48	0.80	0.94
	Arts	0.24	0.27	0.36	0.43	0.60	0.71
Residential	Halls – Residence	0.31	0.36	0.86	1.04	1.17	1.40
	Self-catering/Flats	0.16	0.19	0.72	0.86	0.88	1.06
Libraries	Library – Air Conditioned	1.05	1.45	0.62	0.88	1.67	2.34
	Library – Naturally Ventilated	0.17	0.23	0.41	0.58	0.58	0.81
Students' Union	Students' Union	0.48	0.71	0.48	0.71	0.95	1.43
Administration	Air Conditioned	0.41	0.63	0.32	0.71	0.73	1.34
	Naturally Ventilated	0.14	0.20	0.21	0.44	0.35	0.64
Recreation	Sports/Wet	0.63	0.87	0.95	1.43	1.58	2.29
	Sports/Dry	0.28	0.32	0.63	0.95	0.91	1.27

Source: HEFCE et al. (1996c).

whole system – such as when more efficient windows reduce lighting needs, lead to lower passive heat generation and consequently reduce air conditioning loads (Lovins and Lovins, 1997). Real-world examples back up these claims, but do not indicate the extent or distribution of such opportunities (Romm, 1999). The ultimate potential will, of course, depend upon the nature, role and importance of different barriers.

Table 4.3 Estimates of the energy efficiency potential in the UK service sector to 2010

Measure	**Technical potential (%)**	**Economic potential (%)**
25 costed energy saving measures	30.9	17.8
Other office equipment measures	0.6	0.3
CHP	2.4	2.4
Housekeeping	4.7	3.8
Total	**38.7**	**24.4**

Notes: Economic potential assumes 8 per cent rate of return for costed measures. CHP figure refers to economic potential in both columns.

Source: Fisher et al. (1998).

2.3 Energy and Environmental Initiatives

The 1993 report 'Environmental responsibility: an agenda for further and higher education' was a significant step forward for the UK HE sector (Toyne, 1993). While it was primarily concerned with greening the curriculum, the report recommended that universities adopt comprehensive environmental policy statements, including timetables, targets and the allocation of responsibilities for implementation. Additional recommendations included integrating environmental considerations into capital programmes and adopting environmental performance indicators.

The report provided recommendations rather than requirements, and a follow-up review in 1996 suggested that little progress had been made (Ali Khan, 1996). Thirty-nine institutions had adopted an environmental policy, 21 had appointed an environmental co-ordinator and 15 had an environmental budget. But policies rarely translated to significant action, many co-ordinator posts involved only a few hours a week and the funds made available by most institutions were very small.

The environmental agenda has subsequently been promoted by 'Forum for the Future' (an NGO), which has developed best-practice material in a range of areas, including environmental management systems (EMS) (Forum for the Future, 1997). But progress is slow: universities are sceptical of the costs and benefits of EMS and no university has become accredited. Furthermore, the impact of environmental initiatives on energy management appears to have been minimal. Energy is not perceived as a priority environmental issue and most environmental co-ordinators have little knowledge of, interest in, or influence on energy decisions.

Initiatives to promote improved energy management in the HE sector have been rather more successful. The HEFCs funded a study of energy management in 1996, which led to a range of practical tools and guidance (HEFCE et al., 1996a; 1996b; 1996c). This concluded that savings of 10–20 per cent were possible, but that these would only be realised if there was commitment from senior management and adequate staff resources. There was a wide range of good practice within the sector, and experience suggested that energy management could be financially self-supporting, with savings from utility purchasing and energy efficiency investments being used to fund both future investments and the salary costs of energy management staff. But performance in the great majority of institutions was relatively poor, and responsibility for energy management was ad hoc and unstructured. To overcome this, an integrated approach was recommended, including a formally approved energy policy, structured levels of authority for energy management, and dedicated resources. The study provided detailed plans, geared to university organisational arrangements, that would allow this to be achieved.

The HEFC initiative has provided a range of material that can guide energy management in HE institutions (HEFCE et al., 1996a). But research for the present project strongly suggests that the majority of institutions are not using this material and that very few conform to the best-practice recommendations.

3. UK UNIVERSITY ENERGY MANAGEMENT SURVEY

A postal survey was carried out on energy management practices, technology adoption and perceived barriers to energy efficiency in UK universities in 1999. The survey was sent to the individual responsible for energy management at 94 UK universities,[9] from which 29 valid replies were obtained representing 31 per cent of the survey population. These were classified by total annual energy expenditure, as follows:

Small:	< £1 million/year	11 respondents
Medium:	£1-2 million/year	9 respondents
Large:	> £2 million/year	9 respondents

Data was not collected on floor area as trials had suggested that this could not be reliably obtained. This meant that figures for specific energy consumption (GJ/m^2) could not be derived. The diversity of building sizes, ages and types in universities (Table 4.2) suggests that such data would be of limited value without more detailed surveys. Performance measures are instead derived from self-assessment of variables such as technology adoption. This is subjective and indirect, but nevertheless useful.

The results are summarised below in terms of the percentage of valid replies to each question.[10] The small sample size must be borne in mind when interpreting results expressed as percentages, while the fact that the respondents were self-selecting is likely to have biased the results towards the more proactive institutions.[11] Nevertheless, the results provide a useful picture of energy management practices in one-third of UK universities in 1999.

3.1 Self-assessment Matrix

Respondents were asked to complete the self-assessment matrix on energy management, reproduced in Annex A (section 6), which grades organisational performance from 0 (poor) to 4 (excellent) in six categories (EEO, 1993). Table 4.4 summarises the results in terms of overall weighted average scores. The average score in each category corresponds to:

1. *Policy*: unadopted energy policy;
2. *Organisation*: energy manager, but with authority unclear;
3. *Communication*: contact with users through an ad hoc committee;
4. *Information*: monitoring and targeting reports based on supply meter data;
5. *Awareness*: some ad hoc staff awareness training;
6. *Investment*: short-term payback criteria.

The results suggest a gap between policy statements and initiatives undertaken by estates departments on the one hand, and the general level of communication and energy awareness in the organisations on the other. Self-assessed performance is also correlated with the size of the institution.

Table 4.4 Self-assessment matrix: overall weighted average scores

Category	Small	Medium	Large	Total
Policy	1.8	2.8	3.1	2.5
Organisation	1.4	2.4	2.4	2.0
Communication	1.0	1.9	2.1	1.6
Information systems	1.8	2.1	2.6	2.2
Awareness	1.2	2.0	1.7	1.6
Investment	2.1	2.1	1.9	2.1
Overall	**1.5**	**2.2**	**2.3**	**2.0**

Note: Score from 0 (poor) to 4 (excellent). See Annex A, section 6 for details.

3.2 Energy Policy and Management

A total of 81 per cent of institutions had an energy policy of some form, although this fell to 64 per cent in the small category (Table 4.5). Forty-six per cent of institutions were signatories to the government's *Making a Corporate Commitment* (MACC), which requires signatories to publish a formal energy policy and set targets for energy use.[12] Integration of policy objectives into new build and refurbishment specifications was high at 93 per cent overall (100 per cent for large institutions), while integration into maintenance policy (71 per cent) and purchasing policy (54 per cent) was poorer. 'Integration' here is a subjective concept and the effectiveness of such integration is considered further in the case studies.

An average of 72 per cent of institutions had an energy manager, but fewer than 50 per cent of the small institutions (Table 4.6). On average, 1.4 staff had responsibility for energy management, with the respective figures for the individual size categories being 0.5 (small), 1.6 (medium) and 2.3 (large). Only 39 per cent of institutions had an energy committee, while 58 per cent had produced an annual energy report and 83 per cent had conducted energy audits in major buildings. Energy management was centralised in the estates departments, with only 22 per cent of institutions delegating responsibility for energy to building managers.

3.3 Environmental Policy and Management

No institutions were accredited to a recognised EMS, although 60 per cent stated that they intended to achieve accreditation in the future (Table 4.7). An average of 59 per cent of institutions had adopted an environmental policy, 40 per cent had an environmental committee but only 16 per cent had an environmental manager. Only 13 per cent of institutions felt that the university's environmental policy had made a significant impact on energy management.

Table 4.5 Energy policy (per cent of valid replies)

Question	Small	Medium	Large	Total
Formal energy policy	64	89	89	81
Energy policy objectives integrated into new build and refurbishment specifications	89	89	100	93
Energy policy objectives integrated into maintenance policy	78	71	67	71
Energy policy objectives integrated into purchasing policy	50	25	78	54
Signatory to MAAC	22	56	63	46

Table 4.6 Energy management (per cent of valid replies)

Question	Small	Medium	Large	Total
Energy manager	45	78	89	72
Energy committee	18	33	63	39
Responsibility for energy delegated to building managers	18	33	11	22
Energy audits in major buildings	90	86	78	83
Annual energy report produced	40	56	67	58

Table 4.7 Environmental policy and management (per cent valid replies)

Question	Small	Medium	Large	Total
Environmental policy	45	67	67	59
Environmental manager	18	0	33	16
Environmental committee	60	22	38	40
Environmental policy had a significant impact on energy management	0	22	22	13
Accredited environmental management system	0	0	0	0
Intend to develop such a system	83	25	75	60

3.4 Energy Information Systems

Sub-metering was well established for electricity, with 72 per cent metering at the building level and 19 per cent at the departmental level (Table 4.8). The latter can be problematic if several departments are contained in a single building and/or frequent changes are made in building use, so this may not be feasible for all institutions. Sub-metering for fuel/heat consumption is more problematic and depends upon whether each building has its own boiler or whether there is a district heating system fed from a central boiler. An average of 69 per cent of institutions metered fuel/heat at the building level, with the remainder metering at the site level.

Table 4.8 Level at which energy use is metered (per cent of valid replies)

Energy		Small	Medium	Large	Total
Electricity	Site	28	0	0	9
	Building	64	78	67	72
	Departmental	9	22	33	19
Fuel/heat	Site	50	14	33	31
	Building	50	86	67	69
	Departmental	0	0	0	0

Electricity consumption was recorded monthly by 78 per cent of institutions, while 9 per cent recorded weekly and 13 per cent daily (Table 4.9). The frequency of fuel/heat recording was similar.

Table 4.9 Frequency that energy use is recorded (per cent of valid replies)

Energy		Small	Medium	Large	Total
Electricity	Monthly	82	67	78	78
	Weekly	18	11	0	9
	Daily	0	22	22	13
Fuel/heat	Monthly	89	71	50	76
	Weekly	11	14	17	12
	Daily	0	14	33	12

Table 4.10 indicates how energy data was analysed and used. Eighty-six per cent of institutions monitored trends in energy consumption, with 71 per cent of these adjusting figures for weather conditions. Sixty-two per cent of institutions stated that they used Monitoring and Targeting (M&T)[13] schemes, with the proportion varying from 45 per cent in the small category to 78 per

cent in the large. The HEFC benchmarks were used by 62 per cent of institutions, while a slightly smaller proportion (57 per cent) compared energy use to targets for major buildings. Forty-one per cent of institutions informed building users of their energy consumption while only 29 per cent charged for consumption. The incentives this creates are discussed further in the case studies.

Table 4.10 Energy information systems (per cent of valid replies)

Question	Small	Medium	Large	Total
Monitor trends in energy consumption	73	100	89	86
Adjust records for weather conditions	50	100	67	71
Use monitoring and targeting scheme	45	67	78	62
Compare energy use to targets for major buildings	30	56	89	57
Advise building users of energy performance	20	44	63	41
Charge departments for energy consumption	10	33	44	29
Compare consumption with sector benchmarks	27	67	100	62

3.5 Investment in Energy Efficiency

Only half of the respondents gave a quantitative response to a question on investment criteria for energy efficiency. These reported a mean payback of 4.1 years, with no institutions using internal rate of return (IRR) or net present value (NPV). Only 39 per cent of institutions had an annual budget for energy efficiency investment (Table 4.11), implying that, in the remainder, energy efficiency improvements were achieved through new build, refurbishment or routine maintenance. Of those with a budget, the average ratio of this to the annual energy bill was 8 per cent, although this varied widely from 1 per cent to 19 per cent.

Some 27 per cent of institutions (and 50 per cent of large institutions) stated that they reinvested energy cost savings in future projects. The difficulty here is that energy savings from efficiency improvements may be outweighed by other changes, such as increasing demand from computers and other equipment, and thereby hard to isolate. The only unambiguous measure is the annual bill, but this is influenced by a multitude of factors.

Table 4.11 Energy efficiency investment (per cent of valid replies)

Category	Small	Medium	Large	Total
Use of annual budget for efficiency investment	30	44	44	39
Energy cost savings reinvested to fund future projects	22	11	50	27

Participants were asked how much they agreed that 'a wide range of energy efficiency measures could be implemented that would yield paybacks of <4 years at current energy prices'. The results (Table 4.12) show that 79 per cent of institutions either agreed or strongly agreed with this statement, with 14 per cent neutral and only 7 per cent disagreeing. The potential for energy efficiency improvement was therefore widely acknowledged.

Table 4.12 Wide range of efficiency opportunities with <4 years payback (per cent of valid replies)

Category	Small	Medium	Large	Total
Strongly agree		22	56	38
Agree	36	68	22	41
Neutral	9	11	22	14
Disagree	18	0	0	7
Strongly disagree	0	0	0	0
Don't know	0	0	0	0

3.6 Technology Adoption

Participants were asked to rate the extent to which they had adopted a range of energy efficient technologies, including those most commonly mentioned in best-practice literature.[14] The cost-effectiveness of each technology will depend upon individual circumstances and whether replacement or retrofit is being considered. However, most have been demonstrated to give paybacks of less than five years in common applications (for example, Mortimer et al., 1998).

Table 4.13 summarises the results, with the technologies being listed in descending order of adoption within each subcategory. The following observations may be made:

- Building energy management systems (BEMs) had the widest reported take-up, being extensively implemented by over half of the institutions. But integration of lighting controls into BEMs was rare.
- Widely adopted technologies included high frequency electronic ballasts, pipe insulation, compact fluorescents and 28mm fluorescents. Technologies with limited take-up included condensing boilers, CHP, draught proofing and high efficiency office equipment.
- Small institutions were more likely to have neglected energy efficient technologies. On average, the technologies listed had not been implemented by 19 per cent of small institutions compared to 6 per cent of large.

Statistical tests were conducted to see if there was any correlation between technology adoption, annual energy costs and the performance self-assessment score. In addition, *t*-tests were conducted to identify any relationship between technology adoption and a range of independent variables, such as the existence of an energy policy. In no cases was a statistically significant relationship found. This suggests small or non-existent effects, but the small sample size and the subjective nature of most of the variables presents a serious limitation. In addition, a wide range of factors, such as the technical scope for retrofits, may act to offset any observed relationship.

3.7 Barriers to Energy Efficiency

Participants were asked to rate the importance of 21 suggested barriers to energy efficiency using a scale from 1 (not important) to 5 (very important). Each of the 21 barriers was phrased in terms that would be meaningful to the survey participants, but most related to one of the six more general barriers listed earlier in Table 2.5. The results are summarised in Table 4.14. The overall mean score was 3.3, with the average for all sites ranging from 2.4 to 4.3. Very few respondents rated any of the barriers as not important.

Table 4.13 Adoption of energy efficient technologies/techniques at UK universities

Area	Technology/technique	Mean score
Heating	Use of Building Energy Management System (BEMs)	4.2
	Programming heating and ventilation controls to match occupancy and/or temperature	4.1
	Ensuring thermostats and temperature sensors are located in the correct place	3.4
	Installation of thermostatic radiator valves	3.3
	Improvements to the zoning of heating areas	3.0
	Restricting use of portable electric heaters	2.7
Lighting	Use of compact fluorescents	4.1
	Replacement of 38mm fluorescents with 26mm	3.7
	Use of high frequency electronic ballasts	3.5
	Use of photocell, acoustic or movement sensors	3.0
	Lighting controls integrated into BEMs	1.8
Plant	Use of boiler sequencing controls	4.2
	Insulation of pipes and valves	4.0
	Replacement of oversized boiler plant	2.8
	Replacement of central generation of hot water with point of use applications	2.6
	Installation of condensing boilers	2.5
	Installation of CHP	2.2
Fabric	Fitting door closers to external doors	3.3
	Installation of secondary or double glazing on refurbishment	3.2
	Specification of insulation standards in new buildings that exceed the building regulations	3.2
	Retrofitting insulation to walls and roofs	2.7
	Draught proofing of windows and doors	2.6
Electrical	Power factor correction	3.4
	Use of variable speed drives (VSD) in pumps, fans and other applications	2.9
	Automatic switch off of fans and pumps	2.8
	Specification of high efficiency motors on motor replacement	2.7
	Specification of high efficiency office equipment	2.5
	Specification of high efficiency catering equipment	2.4
Mean		**3.0**

Note: Score from 1 (not implemented) to 5 (extensively implemented).

Table 4.14 Perceptions of barriers to energy efficiency in UK universities

Proposed Barrier	**Total**
Other priorities for capital investment	4.3
Strict adherence to capital budgets	4.0
Lack of capital	3.9
Lack of time/other priorities	3.8
Departments not accountable for energy costs	3.7
Low priority given to energy management	3.7
Conflicts of interest within the organisation	3.6
Energy objectives not integrated into operating, maintenance or purchasing procedures	3.5
Low priority given to environmental performance	3.5
Energy manager lacks influence	3.5
Lack of staff awareness	3.4
Cost of disruptions /hassle /inconvenience	3.2
Poor technology performance (e.g. unreliable)	3.2
Lack of information/poor quality information on energy efficiency opportunities	3.0
Technology inappropriate at this site	2.8
Cost of identifying opportunities, analysing cost effectiveness and tendering	2.8
Cost of staff replacement, retirement, retraining	2.8
Technical risk	2.8
Difficulty/cost of obtaining information on the energy consumption of purchased equipment	2.8
Business/market uncertainty	2.6
Lack of technical skills	2.4
Mean score	**3.3**

Note: Scoring from 1 (not important) to 5 (very important).

Problems related to capital availability were considered the most significant. 'Other priorities for capital investment' scored 4.3, followed by 'strict adherence to capital budgets' (4.0) and 'lack of capital' (3.9). 'Lack of time/other priorities' was the next most commonly reported obstacle (3.8), closely followed by 'departments not accountable for energy costs' (3.7) and 'low priority given to energy management' (3.7). Time constraints may result in large part from the costs associated with employing skilled energy management staff and hence may be an indication of the importance of 'hidden' costs.

Table 4.15 Summary of UK university case studies

Feature	**A**	**B**	**C**	**D**	**E**	**F**
Old/new	New	New	Old	Old	Old	Old
Type	Metropolitan	Metropolitan	Campus	Metropolitan	Campus	Metropolitan
Annual income	£67m	£66.2m	£76.6m	£126m	£187m	£255m
Energy spend	£0.79m	£0.85m	£1.03m	£2.02m	£2.4m	£3.09m
Energy spend as % of income	1.2%	1.2%	1.4%	1.6%	1.3%	1.2%
Size category	Small	Small	Medium	Large	Large	Large
Heat/power ratio	1.9	3.0	2.3	1.8	n/a	1.6
Building mix	22 buildings on 3 sites. Old estate poor quality. New build best practice	>100 buildings on four sites. Diverse ages and many poor quality	~150 buildings. All post 1961	Main buildings are period listed	660 buildings ranging from period listed to recent best practice	300 buildings in 3 clusters Many period listed
Average efficiency (GJ/m^2)	n/a (new build $0.71GJ/m^2$)	$1.29GJ/m^2$	$1.22GJ/m^2$	$0.98GJ/m^2$	n/a	$1.14GJ/m^2$
Self-assessment score	1.5	2.5	2.0	2.2	1.5	2.8
Energy manager	Yes	Yes	No	No (3 staff involved)	Yes	Yes (4 staff involved)
Energy policy	No	Proposed	Proposed	Yes (1992)	No	Yes (1989)

MACC	No	No	No	Yes	No	Yes
Energy targets	No	No	Proposed	No	No	Yes
Energy audits	No	No	No	Yes	No	Yes
Environmental policy	Proposed	Yes	Yes	Yes	Yes	Yes
Payback criteria	3 yrs	3 yrs	3 yrs	3 yrs	Varies	4 yrs
M&T	Yes (1999) Estimated data	Yes (1998)	Yes (1997) Inaccurate and poor quality	Yes (1992) Best practice	Yes (1995)	Yes (1989)
Adjust for degree days	No	Yes	No	Yes	Yes	Yes
Comparison with sector benchmarks	No	Yes (HEFCE)	Yes (limited)	Yes (HEFCE)	No	Yes (limited)
Buildings with BEMS	New	Major and new	One only	All	All	All
Departmental accountability	Space charging	In general overheads	Electricity only	In general overheads	In general overheads	Space charging
Efficiency budget	No	No	5% utilities spend	5% utilities spend	0.4% utilities spend	5% utilities spend

Barriers related to risk were rated relatively unimportant. 'Business/market uncertainty' had an overall score of 2.6, while technical risk scored 2.8. But the respondents (that is, estates staff) may not be best placed to judge the former, and the consequences of business uncertainty may be manifest largely in the strict payback criteria required by management. Generally, these quantitative results provide a useful supplement to case studies and will be referred to subsequently.

4. CASE STUDIES OF ENERGY MANAGEMENT IN UK UNIVERSITIES

Case studies were undertaken of energy management practices within six universities. These included a mix of 'new' and 'old' universities, which in turn were located on a mix of campus and metropolitan sites. Annual energy expenditure ranged from £1 million to £3.1 million. In each case, up to eight people, including the energy manager or equivalent, were interviewed and supplementary information was obtained from the energy management questionnaire and university documents. The case studies are coded A to E, in what follows, in ascending order of annual energy expenditure. Table 4.15 summarises the main features of each institution.

The following summarises energy management practices at the case study universities.

4.1 Organisation

Energy management was typically the responsibility of mechanical and electrical engineers located in the estates department, while separate divisions were responsible for the maintenance of building fabric and for new build and refurbishment. Most projects were carried out by independent contractors, but with project management in-house. In many cases, residences, catering and other areas operated as independent businesses, while university C had building managers in individual buildings. Integrating energy efficiency considerations into all the above areas was a major challenge.

Investment in energy efficiency via a dedicated budget was very much a secondary activity, with the majority of improvements being obtained through routine maintenance, new build and refurbishment.[15] However, investments undertaken via the dedicated budget generally proved to be highly cost-effective. In most cases there was a surplus of potential projects, with the primary constraint being staff time:

'If I looked at all the opportunities with four year paybacks I would need another four staff.'[16]

Energy management staff devoted most of their time to maintaining information and control systems, and to energy purchasing. Time constraints were universally reported as very severe, in part as a consequence of reductions in university 'overhead' staff during the 1990s.

4.2 Energy and Environmental Policy

Only two institutions (D and F) had an established energy policy, and these were the most progressive. Only university F had a quantitative target for energy consumption, and had achieved a 12 per cent reduction in the previous eight years. But despite its considerable efforts in energy management (Box 4.1), electricity consumption had increased by 12 per cent over the same period. All institutions reported increasing electricity use and most were reluctant to adopt either absolute or relative targets for energy consumption in a period when floor area, student numbers, hours of building use, research activity and (particularly) computer use were increasing. The biggest energy savings had been achieved in space heating through the use of improved controls, while falling energy prices and competitive tendering had reduced energy bills in real terms (for example, by 18 per cent in university F over eight years).

Box 4.1 Energy policy at university F

- target to improve energy efficiency (GJ/m^2) by 10 per cent in five years;
- establishment of an improved M&T system;
- investment of 5 per cent of the annual energy budget into cost effective energy efficiency improvements;
- programme of energy audits of the university estate, leading to prioritised energy efficiency investment;
- review of maintenance work against technical specifications and an energy efficiency check list;
- development of energy awareness programmes; and
- inclusion of energy efficiency requirement in the brief to architects and consultants for major building projects.

Five out of the six universities had an environmental policy, and in three cases (C, B and F) this had led to real achievements in areas such as waste management, transport and university grounds. Achievements were less obvious for universities D and E, where the policy was dismissed as 'window dressing'. None of the institutions intended to accredit to a recognised EMS, as the resource requirements were considered to substantially exceed the economic and environmental benefits.

Environmental policy had only a marginal impact on energy management, with the latter being neglected by environmental committees and staff.[17] Even where energy and environmental staff collaborated (university F), energy efficiency measures were solely driven by cost considerations.

4.3 Energy Information Systems and Controls

Information systems varied from rudimentary (C and A) to best practice (D and F). This was a consequence of previous management decisions, and had little to do with the physical characteristics of each site. Establishment of effective M&T systems had been shown to lead to substantial savings. The system at university D allowed comparison with benchmarks, adjustment for weather conditions and occupancy, analysis of trends, identification of problems and priority areas, and checking of utility bills. The savings achieved with this system repaid the capital costs within six months. At the other end of the scale, universities C and A were plagued with inaccurate or estimated meter readings, had practically no information on specific consumption or trends, were unable to identify the savings from energy efficiency projects, and could provide little information to building users on consumption patterns.

The same pattern was evident with building controls. At four institutions, the installation of standardised BEMSs had evolved in parallel with M&T systems and was responsible for the bulk of energy efficiency improvements. These systems allowed centralised control of HVAC throughout the campus according to occupancy patterns and weather conditions. Lighting controls were beginning to be integrated into the BEMS during major refurbishment, with occupancy sensors being installed in major lecture theatres. The overhead costs of running the BEMS were high: universities D and F employed full-time staff; the energy manager at university E spent the bulk of his time running the BEMS; and the non-salary costs for maintaining and upgrading the systems were estimated to be as much as £50k/year.[18] But these costs were considered to be more than recovered by savings in energy costs. In contrast, university C provided a salutary example of the perils of non-standardisation:

'At the moment there is a whole hodge podge of different controls...We had a policy of selecting the lowest cost tender rather than standardising on one system. Often we'd end up with a system which was completely different from the last project...It creates all sorts of problems with spares and maintenance and knowing how to use the equipment...'[19]

Here, a combination of inadequate controls and poor zoning had led to considerable inefficiencies:

'This building is ridiculously overheated. Some people on one side of the building may be OK while those on the other side are roasting.'[20]

'...we don't have information on energy use, hours of occupancy, or on what people want ... This means we could be heating the building at weekends when no-one is there because nobody has reset the time clock ... You only get to hear about it when someone complains.'[21]

Most institutions reported rising electricity consumption on both a relative and absolute (GJ/m^2) basis. But inadequate monitoring and record keeping meant that consumption trends could not be easily identified and compared.

4.4 Accountability and Incentives

Only university C directly charged departments for electricity use, while none charged for fuel/heat. Instead, utilities costs were included within general overheads, with three institutions linking this to space occupied. The only exception to this rule were on-site businesses such as catering and shops. These were frequently charged directly for utilities, since a combination of their overall business orientation and relatively high energy costs made them more responsive to price incentives than academic departments. For the latter, there were three serious obstacles to departmental charging:

- inadequate sub-metering, coupled with scepticism that the cost of additional metering would be recovered from either improved housekeeping or investment in energy efficiency;
- shared occupancy of individual buildings, combined with frequent changes of room use and departmental location; and
- political sensitivities associated with increasing costs to energy intensive science and engineering departments.[22]

Scepticism over the effectiveness of departmental charging was borne out by the experience of university C, where departments were charged for electricity consumption. The arts, humanities and social science departments had no technical capacity to improve energy efficiency and simply paid the bill. Science departments had more capacity through the use of technically competent building managers, but their room for manoeuvre was severely limited by rigid capital budgeting rules and the dominance of research priorities. Meanwhile, the incentives for the estates department to fund (electrical) efficiency improvements were blunted as the host department kept the savings. Shared savings schemes had not been considered, due to the transaction costs involved. As a consequence, the system of 'devolved budgeting' was cited as an *obstacle* to improved energy efficiency. In one department, a heat recovery project with a 15 month payback had been delayed for several years owing to lack of funds in the devolved capital budget.

4.5 Capital Budgeting and Investment Criteria

Three institutions had a dedicated budget for energy efficiency investment equivalent to 5 per cent of the annual utilities bill, while a fourth had a token £10k/year (0.4 per cent). But the bulk of energy efficiency improvement derived from specifying energy efficient technologies (for example, high frequency lighting) as standard in new build and refurbishment. The dedicated energy efficiency budget was used for a combination of stand-alone investments and top-up funding of maintenance to ensure cost-effective options were chosen. Constraints on staff time meant that no institution had developed a portfolio of potential projects, and some had difficulty in spending the existing budget within the financial year. But in all cases, the investment opportunities appeared considerable. For example, all recent projects funded at university D had a payback of less than two years, and the energy manager at university E had developed outline plans for an investment programme of £2 million over five years (~16 per cent of energy costs/year), all for projects with four year paybacks.

The rigidity of capital budgeting was a common theme and made it difficult to trade-off capital expenditure in the current financial year with long-term revenue savings. This led to many opportunities being missed:

> 'The accounting system constrains you...If you go into deficit in one year you are in trouble, even if it would pay back next year...You can't overspend, even if it means lower life cycle cost...The capital budget is completely separate from the revenue budget. So higher capital cost just shows as a higher capital cost – you can't trade-off between the two...'[23]

Paybacks were preferred to IRR or NPV on the grounds of simplicity and convenience. Four institutions used three year payback criteria, while university F required four years. The rationale for stringent paybacks was never explicit, with interviewees referring instead to 'sensible' paybacks' or 'good practice':

> 'Three year payback is just my gut feel. It is not as if I have discussed this with anybody and established a policy…There is no guidance that I am aware of…it's not something I have any big argument for or against…'[24]

4.6 New Build and Refurbishment

All institutions had major ongoing capital programmes, funded from a wide range of sources and including public-private partnerships for residential developments (Grout, 1997). Two institutions (E and A) had attempted to deliver best practice in energy and environmental performance in several major projects, and had achieved considerable success. Nevertheless, all institutions reported a range of barriers within the construction process which acted to undermine energy efficiency. These problems derived primarily from the structure and organisation of the UK construction industry, including the 'linear' rather than integrated design process, the reliance placed on cost-based competitive tendering and the scope for opportunistic behaviour by individual contractors.

These problems are complex and far reaching in their implications, and are explored in more detail in Chapter 7. But three factors 'internal' to universities are also relevant. First, and most importantly, there were very strong pressures to keep within the capital budget for major projects, with the result that cost-effective energy efficiency features were frequently dropped:

> 'The original estimate [for a new building] was £4 million, but after the details had been worked out we had an estimate of £6 million. The instructions were "get the price down, it's too much". So a lot of energy management things were thrown out…Some things that were removed had a very good rate of return. The driver was get the project within the budget...with a neglect of the future revenue consequences.'[25]

Second, there was a lack of integration of energy objectives into the building procurement process, including the absence of targets for energy and environmental performance and inadequate guidance to designers and contractors in areas such as zoning and BEMS. While the delivery of energy efficient buildings does not lend itself to prescription, the relative success of university F rested in part on such integration, while its absence elsewhere had created serious problems. Finally, there was a lack of consultation between designers, contractors, estates staff and building users, leading to problems such as incompatible control systems and inadequate user control.

4.7 Purchasing and Policy Integration

Each institution operated a system of devolved financial management in which the authority to purchase goods rested with departments and individual members of staff. Each had a purchasing department and corresponding purchasing procedures, but the formal involvement of this department was only required for purchases above a certain cost threshold (for example, £15k). Larger purchases were subject to guidelines and regulations from the EU, the UK government and the HEFCs, with the latter emphasising value for money through the use of whole life costing (WLC).[26] In addition, several institutions used regional university purchasing consortia to obtain price reductions through bulk purchase and other benefits.

While lifetime energy costs are significant for many items of equipment (for example, photocopiers, fume cupboards), energy appeared to be neglected or ignored in the majority of purchasing decisions.[27] Similarly, whole life costing was the exception rather than the rule within both individual institutions and the purchasing consortia.[28] The neglect of energy considerations had important consequences. For example, only one of the six institutions required the purchase of Energy Star compliant computers, and here the power management facilities had never been enabled – despite potential annual savings of £50–100k.[29] Four institutions failed to require energy efficient motors on motor replacement, and at one institution the maintenance department were still using 38mm rather than 26mm fluorescents, despite the latter being 10 per cent more efficient for the same purchase price.[30] Liaison between energy management staff and the purchasing department was absent or rare, and equipment purchasers rarely consulted energy management staff:

> 'Departments are required to consult us, but they don't…A typical example is where they buy something that needs water or cooling. The cheapest way is to put the water down the drain. But a recirculating unit can pay for itself within 1.5 months. Departments don't want to pay the initial cost. Even when we offer to pay a contribution towards it, they are still reluctant.'[31]

Researchers responsible for equipment purchases were accountable for capital costs but not running costs, lacked information on and knowledge of energy issues, and were primarily concerned with research needs. Similarly, maintenance staff were frequently engaged in 'firefighting' rather than planned maintenance and were primarily concerned with restoring full use of equipment as soon as possible. In such contexts, neglect of energy efficiency becomes inevitable. While there were moves to improve guidance on WLC and other issues, there was a concern that strict guidelines would be seen as violating academic freedom. Energy management staff preferred informal liaison where appropriate, but resource constraints prevented them from becoming involved in the majority of purchasing decisions. 'Micro-management' of issues such as the choice of new catering equipment was seen to be neither feasible or appropriate.

4.8 Awareness and Culture

The level of energy awareness among staff and students was uniformly reported to be very poor. As a consequence, the general level of housekeeping (for example, turning lights and computers off) was also poor. Factors contributing to this included the relative insignificance of energy costs, the lack of accountability for energy costs and the dominance of research and other priorities. While a number of institutions had undertaken limited awareness campaigns (for example, stickers, exhibitions) these were generally considered to be ineffective. The environmental consequences of energy consumption were considered to be of little help in motivating staff, owing to their relative invisibility compared to problems such as waste. Partly as a consequence of this, estates departments preferred to concentrate on improving energy efficiency through better controls, rather than improved housekeeping.

In most cases, the knowledge and commitment of top management was also limited:

> 'You are not going to get anywhere if the guy up top isn't clued up. They don't believe the message that if they put their hands in their pocket, they will save in the long-term.'[32]

This could partly be resolved through clear information to management on energy costs and saving opportunities. But this requires adequate monitoring and staff resources and was not yet feasible at all institutions.

4.9 Energy Services and Outsourcing

University C was considering an energy service contract for a CHP district heating scheme financed through a public-private partnership. The drivers here were the need to replace existing boiler and distribution equipment, coupled with a serious lack of funds for major investment. But the other institutions lacked comparable drivers and were very sceptical of energy service contracts. Concerns related to the difficulty of including and enforcing quality criteria in service contracts, the loyalty of staff to the service company rather than the host institution, the preference to retain cost savings in the host institution, rather than share them with an outside company, and the sensitive internal politics of contracting out – including redundancies and changed conditions of employment. These concerns outweighed the potential benefits of energy service contracts, despite the internal resource constraints identified above. Interviewees gave many anecdotal examples of the risks of energy service contracting:

> '[a recent energy service contract] was a disaster. They were questioning every piece of work...All the work that was normally done by direct labour was treated as additional to what was specified in the contract. Hence it required more money...I don't think you can get round this by writing the contract in a better way. It is a question of attitude as well. You can only be so prescriptive...'[33]

5. BARRIERS TO ENERGY EFFICIENCY IN UK UNIVERSITIES

The results from the survey and case studies confirm the results of independent studies in suggesting that there is a substantial potential for cost-effective energy efficiency improvement in the UK HE sector which is being obstructed by a variety of barriers. Those barriers which appear of greatest importance are listed in Table 4.16 and discussed in turn below.

5.1 Access to Capital

Survey respondents considered lack of capital, in various forms, to be the most important barrier to energy efficiency (Table 4.14), while interviewees highlighted the rigidity of capital budgeting procedures and the serious consequences of capital constraints in major projects. The severe financial difficulties faced by one institution had led to staff cuts, recruitment freezes, and the removal of the energy efficiency budget:

> 'If you're running at a surplus you can afford to invest in energy management. But it doesn't happen here. We focus on paying the bill at the end of the week...Long term strategic decisions get lost...a short term

focus prevails...My big question is how we get out of this downward spiral of structural deficits in the budget...'[34]

There were several aspects to the 'lack of capital' problem. The first was the severe financial constraints faced by the sector. In 2000, HEFCE noted that current levels of funding were insufficient to finance capital expenditure from retained earnings, while it considered that the ability to take on additional borrowing was limited by the capacity to repay from operating cash flows (HEFCE, 2000a). The annualised servicing costs of all long-term financial commitments by the sector averaged 1.7 per cent of total income, which was less than the 4 per cent limit beyond which HEFC approval was required (HEFCE, 2000b). This rule provides a means of safeguarding public money in the context of agency relationships between the HEFCs and individual institutions, but also leads to an aggregate restriction on cost-effective investment. Much will depend on how the rule is interpreted in individual cases, but any application to the HEFCs to exceed the 4 per cent ceiling is likely to entail substantial transaction costs. In practice, however, the dominant factor restricting borrowing appeared to be management concern over financial prudence, which was influenced by both the HEFCE rules and sector average debt/turnover ratios.

There is clearly a disjunction between the theoretical ability to borrow for low-risk, cost-effective energy efficiency projects, and the risk associated with increasing aggregate borrowing by an institution that is only just breaking even. While small-scale energy efficiency investments should have little impact on the level of debt for the university as a whole, borrowing may not be considered for minor projects due to the transaction costs involved and it will generally be difficult or impossible to 'ring-fence' the revenue stream of avoided energy costs. Restrictions on aggregate borrowing may therefore restrict the budget for energy efficiency investment, despite the apparently high returns.

It was notable that none of the case study institutions had considered external financing of energy efficiency investment as this was outside either the responsibility or competence of the majority of staff interviewed. The staff who could identify possible projects were preoccupied with working to existing budgets, subject to severe time constraints, unused to making a business case for additional finance, and constrained by university

Table 4.16 Barriers to energy efficiency in the UK higher education sector

General category	Specific instance	Comments
Access to capital	Availability of capital to the university	Funding Council imposes aggregate restriction on borrowing, while borrowing is rarely contemplated for small energy efficiency investments. Capital constraints in funding for new construction projects are a major source of inefficiency.
	Allocation of capital within universities	Research investment has greatest priority. Budgets for energy efficiency investments are typically less than best-practice recommendations. Capital budgeting procedures are rigid and inflexible.
Hidden costs	Overhead costs of energy management	Salary overheads are very significant – typically 60 per cent of the investment budget. This is manifest in severe time constraints on staff. Overheads may need to be recovered in part from energy savings on investment projects.
	Cost of identifying opportunities, analysing cost effectiveness and tendering	Highly significant in the context of severe time constraints on energy management staff.
Imperfect information	Information on organisational energy use	Many institutions could significantly improve their energy information systems. Communication of information to management, staff and students is poor.

Split incentives	Departments not accountable for energy costs	Perceived as a major obstacle. But there are major costs and difficulties associated with introducing devolved budgeting.
	Purchasers of equipment not accountable for energy costs	Limited adoption of whole life costing, few attempts at integrating energy concerns into purchasing criteria and limited consultation between estates and purchasing departments.
Bounded rationality	Decision-making and organisational routines may neglect energy efficiency	Prevalence of severe time constraints and the small contribution of energy to total costs leads to the neglect of energy efficiency in individual and organisational routines. Simple paybacks preferred to rate of return calculations on the grounds of simplicity and convenience.

procedures. In contrast, financial staff often knew little about the possible opportunities and considered borrowing unnecessary for sums as small as 5 per cent of the utilities budget, given the associated transaction costs. At an institution where the energy manager stated that he would need four people to implement all the possible energy efficiency projects, the financial manager commented:

> 'You are talking about saving a few £k per annum. It's peanuts. If we were talking about a £1m pot then it's different. But bring me the projects. They are not coming forward, so I am confident that we have high efficiency here already.'[35]

Capital constraints had their greatest impact in new construction and major refurbishment and the resulting loss of opportunities is documented more fully in Chapter 7. Key issues here included: the low priority of cost saving investments compared to other features such as increased space; the separation of capital and revenue budgets; and the combination of a linear design sequence and frequent cost overruns, leading to energy efficiency features being removed to keep within budget. Again, the university should be able to borrow additional money to fund energy efficiency features with a high rate of return that would otherwise be cut from the project. In practice, university staff may not be made aware of the consequences of design choices until it is too late. Also the transaction costs of raising additional finance to fund specific design features or individual items of equipment are likely to be high, while the incentives on relevant parties to reduce life-cycle costs are much smaller than those to meet the capital budget. There may also be cost control issues, with the perceived risk that additional funds would get lost in the overall capital budget rather than being dedicated to specific cost-effective improvements.

A third feature of the 'access to capital' problem was the absence, or small size, of dedicated budgets for energy efficiency investment. This derived from priority-setting within the institutions, which also explained the relative neglect of building maintenance during periods of financial stringency. Research and teaching took priority as these were viewed as strategic and essential to the future of the institution. In contrast, energy efficiency was discretionary, and its low priority was reinforced by time constraints, lack of knowledge/information, falling energy prices, the small contribution of utilities to total costs and its relative invisibility. While business development investments and new research grants were highly visible on the balance sheets, the money saved on the utilities bill from an energy efficiency investment was not. This was particularly the case if the energy savings were outweighed by changes in energy prices or by energy demand growth elsewhere in the institution.

A final feature was the effective capital rationing created by stringent investment criteria. A hypothesis from agency theory is that these are imposed by management to ensure that only unambiguously cost-effective projects go ahead. In the university context, this seems less persuasive. First, none of the case study institutions had strict guidelines on payback criteria and in four cases the estates department itself chose the criteria. Second, no interviewees were able to give a clear rationale for the use of short paybacks, making instead rather vague references to 'good practice' and uncertainty. Against this, it was notable that neither the financial performance of individual investments nor that of overall energy management was monitored closely. While annual energy costs could provide a performance measure, these would need to be corrected for a range of factors including degree days, changes in energy prices, changes in taxation, increases in student numbers, changes in building use and so on, all of which implies information costs. This may partly explain why the employment contracts for energy managers contained no direct incentives for improving energy efficiency, but here the lack of precedent for such contracts in the sector coupled with staff and union opposition was also relevant.

In sum, from the perspective of survey respondents and case study interviewees, limited access to capital provided a significant barrier to energy efficiency. Explanations for this can be developed with reference to the overall financial constraints faced by the organisations, management priorities, bounded rationality and transaction costs. Nevertheless, the blanket restrictions on borrowing that pervade the sector, coupled with the reluctance to access external finance via energy service contracts, suggest a failure by university management.

5.2 Hidden Costs

Hidden costs may usefully be grouped into three categories: the overhead costs of energy management; costs specific to a technology investment; and the loss of utility associated with an energy efficient technology. Table 4.17 indicates the survey respondents' evaluation of the importance of these costs.

The results suggest that the staff time associated with implementing energy efficiency investments was considered very important, while other types of hidden cost were less important. Each general category is discussed below.

Table 4.17 Importance of hidden costs in the university energy management survey

General category	Postulated barrier	Importance score	Ranking in list of 21 barriers
Overhead	Lack of time/other priorities	3.8	4
Specific	Cost of disruptions, hassle and inconvenience	3.2	12
Utility	Poor technology performance (e.g. unreliable)	3.2	13
Specific	Cost of identifying opportunities, analysing cost effectiveness and tendering	2.8	16
Specific	Cost of staff replacement, retirement and retraining	2.8	17
	Mean	**3.2**	**12**

Note: The average importance score for the 21 postulated barriers was 3.3.

Overhead costs of energy management

Time constraints on energy management staff are a good indicator of the importance of salary overhead costs. Both survey respondents and interviewees identified time constraints as the most important barrier to energy efficiency and invariably reported that they were getting worse. Opportunities were missed not because of lack of capital, but because of lack of time to identify and implement the investment. Where capital and information barriers existed, they were usually interlinked with time constraints:

> 'The biggest obstacles are money and time. These are linked. We need more time to spend more money and vice versa. We are approaching saturation point at the moment. We could easy take on one more person.'[36]

> 'There is no shortage of things to do. We could easily have another two guys here.'[37]

The importance of salary overheads is vividly illustrated by university C, which did not have a full-time energy manager despite having an annual utilities bill of £1.5 million. Interviewees considered that to obtain funding for a full-time energy manager, they would need to demonstrate how such a person could save his/her salary costs plus 50 per cent over a three year period. Assuming annual salary costs of £30k, this means that the manager would need to make net annual savings of ~£45k, equivalent to 3 per cent of the utilities bill or 64 per cent of the energy efficiency budget. But since the latter was already achieving savings, this means that the *additional* savings from employing a full-time energy manager would need to reach this level. The difficulty of achieving this, together with the lack of staff time to prepare such a case, was preventing the appointment of a full-time manager – despite the serious consequences for energy management.

The overhead costs of energy management at the case study institutions were difficult to quantify since the relevant duties were spread between a number of staff. As a ballpark figure, the best-practice guidelines of a full-time energy manager for each £1m of annual utilities (that is, energy + water) expenditure equates to salary costs of around 3 per cent of the utilities bill (EEO, 1993, p. 19). The only case study institution to come near this was university F, but here the relevant staff spent much of their time on waste management and other activities.

Three institutions (C, D and F) had an annual energy efficiency budget equivalent to 5 per cent of annual utilities expenditure. Following best-practice recommendations, the salary costs of energy management would equal 60 per cent of this budget. Viewed in this way, stringent payback criteria may be justified as a means to recover the salary overheads. But in the case of university C, three years was perceived as the required criteria *including* salary costs, suggesting that an even shorter payback (~18 months) was required for the capital costs alone.

As argued in Chapter 2, much of energy management can be seen as an essential overhead necessary for the functioning of the institution and the comfort of employees. If energy management tasks are neglected due to a lack of staff resources, the result may be higher energy prices, equipment breakdowns or employee discomfort – all of which were evident at university C. From this perspective, the salary overheads of energy management staff need not solely be covered from the savings on energy-specific investment projects. Some may be covered though negotiating lower energy supply tariffs or by maintaining and operating equipment more efficiently, while the wider environmental contribution of energy efficiency may be considered part of the strategic, non-financial goals of the institution (a rationale that had been employed by university C to justify appointing an environmental manager). The same arguments apply to the cost of operating and maintaining energy information systems. While high (~£8k/year or ~25 per cent of salary costs for a full-time manager),[38] some of the biggest savings from such systems result from challenging incorrect utility bills and identifying and rectifying equipment faults soon after they occur.

If core energy management tasks are being adequately performed, it may be justifiable to recover the marginal costs of *additional* staff resources from the savings in energy costs. However, in several of the case studies it is arguable that the core tasks were not being adequately performed. Furthermore, energy management overheads were neither separately quantified or linked to investment criteria. Instead, they were included within overall estates overheads and recovered indirectly through space charging. Also, only two of the case studies had conducted campus-wide energy audits to identify the potential scale of investment opportunities.[39] Hence, the 'hidden cost' mechanism can only operate in an indirect manner, through managerial decisions to reduce (or not expand) staff numbers, based upon inadequate assessment of the costs and benefits.

Costs specific to a technology investment

Technology-specific costs were considered less important than general overheads in the survey, and few interviewees made specific reference to, for example, the cost of identifying opportunities and tendering. But to the extent that these take up management time, they are subject to the constraints indicated above. It was notable that institutions concentrated on large projects with significant savings, rather than numerous small projects, as project management formed a disproportionate share of total costs for smaller investments. Similarly, very few investments from the energy management budget were subject to detailed investment appraisal since they were too small to justify the time and costs involved. Instead, projects went ahead on the basis of a broad knowledge of the cost-effectiveness of a particular technology. But an important theme was the costs associated with gaining approval for investment:

> 'The decision making process is very slow – and not responsive to change. At least two committees are required to get a project authorised…It is not like a commercial business. There are lots of competing aims and complicated organisational politics…We need more accountability, but delegation is not part of the university's culture...'[40]

In the more successful case studies (for example, D and F), the estates department had considerable autonomy and flexibility in spending its budget, but where lengthy approval was required through the university decision-making system, energy efficiency opportunities were missed. An extreme example was university C, where projects with paybacks as short as 15 months were being rejected by building services managers because they were unable to fund them within their departmental budget and because the perceived cost of the application process for additional funds was too high (indeed, some managers were not aware that such a process existed).

Hassle and inconvenience in installing equipment was ranked twelfth in the survey and the case studies suggested this was relatively unimportant. Institutions were less vulnerable to inconvenience than manufacturing; most refurbishment work was done during the summer recess; and most improvements were achieved through refurbishment work which was going ahead anyway, so the hassle could not be attributed to energy efficiency improvements alone.

Loss of utility with energy efficient technologies

Problems associated with loss of utility scored only thirteenth in the survey and the primary example given by interviewees was lighting controls.[41] Such features are specific to individual technologies and it is difficult to generalise from the case study evidence. However, it is notable that reports of additional benefits from energy efficient investments outnumbered the reports of loss of utility. For example, high frequency lighting assisted compliance with VDU regulations; double glazed windows reduced maintenance costs for painting; and an energy efficient alternative to wet humidification avoided problems with biological growth and thereby reduced cleaning costs. While anecdotal, these results suggest that the additional benefits of energy efficiency investments are as least as common as loss of utility for the technologies considered.

Summary

Salary overheads are a very important hidden cost and the resulting time constraints can partially explain the neglect of energy efficiency opportunities. In many cases a bias towards cost-cutting through redundancies appears to have left institutions with insufficient staff. However, it is difficult to quantify these overheads and since universities do not have an explicit requirement to recover overheads from investment projects, hidden costs cannot fully explain the choice of strict investment criteria.

A conflict is also apparent between autonomy and accountability in energy decision-making and difficulties are often created by bureaucratic university procedures. In contrast, the costs associated with disruptions or the loss of utility from energy efficient technologies seem much less important.

5.3 Imperfect Information

Effective energy management requires adequate information on the volume and pattern of energy use in the organisation, the opportunities available for energy efficient investment, and the energy consumption of new and refurbished buildings and purchased equipment. In each case, obtaining, transmitting and using this information entails transaction costs.

The case studies suggest that high quality information systems are a prerequisite for good performance in energy efficiency, and that investment in information systems can be highly cost-effective. But three of the case study institutions and over a third of the survey participants could significantly improve their information systems.[42] Furthermore, the usefulness of information on energy consumption was hampered by the absence of standardised and widely accepted benchmarks for the sector.

In most cases, the communication of energy information to management, staff and students was poor. Senior management appeared almost entirely unaware of either the achievements that had been made or of the opportunities that were available. While better information strengthens the case for energy efficiency investment, obtaining that information itself requires investment:

> 'The crux is demonstration. If you can prove that introducing something saves money then the Head of Finance will support you. You have to monitor at a low level to do this.'[43]

'Lack of information/poor quality information on energy efficiency opportunities' was ranked fourteenth in the survey and interviews suggested that the problem was not so much the availability of information within the relevant energy service markets, but the lack of time to read literature and attend workshops. The survey found that a wide range of information sources on energy efficiency were used, but government best-practice material was considered the most useful. Of particular value was the government-sponsored Higher Education Energy Share Fair network, which combined an email list with regular workshops and used a system of offers and requests to allow energy managers to share problems and experiences with colleagues in other universities. This forum is specifically designed to facilitate trust and encourage mutual learning, which in turn lowers the transaction costs of information acquisition.[44]

The great majority of interviewees considered they had sufficient information on energy efficiency opportunities and pointed to time or capital as the more important constraints. But while lack of information on generic technologies may not be a problem, the paucity of energy audits suggested that they lacked information on specific opportunities. Similar information problems were apparent in equipment purchasing, but here the search costs entailed in obtaining information merely served to reinforce the neglect created by lack of accountability for energy costs and lack of knowledge of, or concern for, energy issues.

5.4 Split Incentives

'Departments not accountable for energy costs' was ranked fifth in the survey, with only 29 per cent of institutions charging departments for energy consumption. Only one of the case studies charged departments for electricity, none charged for fuel/heat, and interviewees emphasised how the lack of accountability undermined incentives for housekeeping or investment. But improving accountability is a difficult and sensitive issue. Only a small number of UK universities have introduced 'devolved budgeting'[45] and it seems unlikely that many more will do so within the next few years. Improving accountability entails costs which will vary according to physical layout, the current extent of sub-metering, political sensitivities and other factors, and the case studies demonstrated how such initiatives may be ineffective or even counterproductive. This suggests that, in many cases, centralised energy management may be more effective than introducing dilute incentives to which (boundedly rational) actors are unable or unwilling to respond.

> 'Academics aren't interested. They don't pay the bill. However, even if they did pay the bill it would be a perfunctory thing – they wouldn't notice. Energy costs are small and their priority is research. The university's position is that devolved budgeting would be an administrative nightmare. I think this is well founded.'[46]

'Energy objectives not integrated into operating, maintenance or purchasing procedures' was ranked sixth in importance in the survey. For purchasing, the case studies suggested limited or ineffective use of whole life costing, the absence of guidelines/requirements on energy efficiency and little consultation between relevant groups. While lack of accountability for running costs was the primary obstacle here, the scope for resolving this is limited both by the difficulties indicated above and by the transaction costs associated with overseeing a large number of small-scale purchasing decisions. This in turn points to the potential role of equipment efficiency standards, labelling schemes and consortia purchasing.

In contrast to purchasing, the integration of energy objectives into routine maintenance appeared to be fairly good, despite the lack of accountability for running costs. Relevant factors here include the knowledge and motivation of maintenance staff, the working relationships between the energy manager and the maintenance manager (typically colleagues in the same division) and the requirements established in university policy and procedures. As discussed below, much also depended on the nature of the maintenance task and the associated time and capital constraints.

In summary, split incentives in a variety of forms appear to be a significant barrier to energy efficiency in UK universities, but overcoming these barriers may in many cases entail transaction costs that exceed the benefits to be achieved.

5.5 Bounded Rationality

Boundedly rational decision-makers will make satisfactory rather than optimal decisions using imprecise routines and rules of thumb. In doing so, the level of consideration given to energy efficiency will vary not just with the economic importance of energy, but also with the competence of the actor, the level of information available, actors' motivation and incentives, the rules and procedures governing their actions, time constraints and the nature of the decision (for example, emergency versus planned replacement). In universities we find severe time constraints at all levels, a dominance of research and teaching priorities, rigid and bureaucratic decision-making structures, complex organisational politics, a growing funding crisis and a frequent neglect of commercial considerations within decision-making. These factors combine with inadequate information on energy use, lack of accountability for energy costs, declining energy prices and the relatively small contribution of energy to total costs to encourage individual and organisational routines that neglect energy efficiency. As the interviews regularly demonstrated, these problems may apply *even if* the individual or department is paying the energy bill and/or is well informed about the consequences of their actions:

> 'When we replace a pump…we are governed by the physical dimensions, the duty, and the cost. Probably no other questions get asked…It's normally an emergency situation. The existing one may be 30 years old and they don't make them any more. We may have flanges with imperial measurements and new ones are metric – etc. etc...It's driven by need to replace it quickly. If one is available, get the same type as what's in at the moment. If not, get something that will physically fit. Energy efficiency just doesn't come into it...'[47]

One way in which case study institutions overcame such constraints was through building energy efficiency into organisational procedures in such a way that little thought or assessment was required. For example, university D always required high frequency lighting in refurbishment or replacement, thereby preventing departments from choosing lighting with a lower initial cost but higher life-cycle costs.[48] However, such integration was relatively patchy and written guidelines often conflicted with other objectives such as the desire for autonomy in decision-making. Conflicts with strict capital budgets were of particular importance and capital constraints generally took precedence over energy efficiency guidelines even when this meant that highly cost-effective investments were forgone. From the perspective of a boundedly rational and time constrained decision-maker who is not accountable for running costs this may be sensible – the transaction costs associated with calculating rates of return and gaining administrative approval for exceeding a capital budget may be too great to be worthwhile

Bounded rationality thereby compounds the neglect of energy efficiency that is encouraged by other market and organisational factors. Some decision-making routines should be amenable to change, such as the encouragement of whole-life costing for large items of equipment. But for others, such as the emergency replacement of motors, energy efficiency is never likely to feature highly. It is in these cases where public policy initiatives such as market transformation schemes may be both appropriate and cost-effective.

6. POLICY IMPLICATIONS

The majority of the barriers described above result from the transaction costs of acquiring, transmitting and using information on energy efficiency, together with the costs of overcoming information asymmetry and split incentives within complex, hierarchical organisations. Since many of these barriers result from organisational goals, structures and decision-making routines, changes at the organisational level are an appropriate response. But these are unlikely to be forthcoming while energy costs and environmental performance remain a marginal concern. While management attention could be raised through carbon taxation, prices are a blunt instrument and need to be supplemented and reinforced by other approaches. This section highlights some possible initiatives at the organisational, sectoral and government level and describes how recent developments in EU and UK climate policy are affecting the HE sector.

6.1 Measures within Individual Universities

There are a range of measures which individual universities could undertake to improve energy management, and there are some excellent best-practice examples which could be followed (Somervell and Talbot, 1991; HEFCE et al., 1996a). These particularly apply to institutions with a utilities bill exceeding £1 million/year, but simplified measures may be adopted by smaller institutions. The costs of establishing effective energy management need to be weighed against the potential benefits as well as the alternative route of energy service contracting. But the creation of an energy management profit centre is particularly attractive as it gives autonomy in decision-making and could allow energy management to become self-funding.

Table 4.18 has been developed from the case study results and best-practice literature (for example, Croner, 2000) and illustrates the basic elements of best practice in university energy management. For such initiatives to be adopted, energy efficiency must become a strategic issue for top management. This is a challenge in the current business situation, and points to the role of national climate policy.

6.2 Measures within the Higher Education Sector

There are a number of measures that could be undertaken by government, the HEFCs or other bodies and targeted specifically at the HE sector. Of these, the development of relevant sector-specific information on energy efficiency opportunities is of particular importance. Best-practice literature, including case studies is already available and the Energy Share Fair Network is a model of effective dissemination. But while the former could be expanded and more effectively targeted and marketed,[49] the biggest weakness is the lack of credible standardised benchmarks for the energy performance of university buildings. The existing HEFC benchmarks provide a starting point, but these are widely criticised and insufficiently used.

The HEFCs have an important role in overseeing financial management and promoting 'value for money'. This role may change as the government seeks new funding sources for HE,[50] but one relevant option would be to review the constraints on university borrowing (HEFCE, 2000b). One possibility would be to remove the 4 per cent threshold altogether, while another would be to relax the approval requirements for investments that generate an income stream (including energy efficiency), but to retain oversight for other investments (for example, lecture theatres) where the income benefits are less direct. Alternatively, the HEFCs could encourage improved energy management through closer attention to estates practices in their audits of individual institutions (HEFCE, 1997).

A more effective way of improving energy information systems in the sector would be to introduce mandatory reporting requirements for energy use, energy intensity and greenhouse gas (GHG) emissions. The HEFCs have recently explored the use of 'key estate ratios', including energy costs/m^2, to facilitate comparison between institutions (HEFCE, 2001), and these could be extended over the longer term to the collection of more disaggregated energy statistics and comparison against benchmarks. Few institutions have the capacity for this at present and the costs and benefits would need to be carefully assessed. But improved energy information should stimulate cost savings through improved energy management.

A logical step from improved reporting would be the adoption of targets for GHG emissions. The UK climate programme states that: '.. public sector buildings outside the government estate should achieve energy efficiency improvements on the same scale as those that are set for the government estate', [51] which are for a 1 per cent/year reduction in GHG emissions up to 2010 (DETR, 2000a, p113).[52] The HEFCs could propose the adoption of GHG emission targets by individual institutions and performance against these targets could be linked to financial incentives within the existing funding system. A precedent here would be the mandatory targets for GHG emissions adopted by UK hospitals and administered through the executive agency responsible for hospital estates (NHS Estates, 2002).[53] While penalties for non-compliance are not explicit, the targets are taken seriously by NHS Trusts.[54] If a significant 'no-regrets' potential exists, such targets could contribute to improving value for money in the HE sector as well as serving an environmental purpose. At present, however, there seems little prospect of such a move.[55]

Improvements in university purchasing procedures can also play a role. The Joint Procurement Policy and Strategy Group (JPPSG) oversees procurement in the HE sector and has published a good-practice guide on whole-life costing, together with software tools (JPPSG, 1998). A 1999 National Audit Office (NAO) report recommended that institutions adopt WLC techniques and put controls in place to ensure that they are correctly used. This recommendation needs to be followed through, and could be encouraged via the HEFC audit procedures. In addition, if regional purchasing consortia could procure or recommend certain items of equipment on the basis of WLC, this would benefit individual universities without requiring internal changes to procedures and routines. A number of consortia have set up working groups on 'green procurement' which could provide a route for ensuring WLC for common purchases in all participating institutions (Forum for the Future, 1999).

A final possibility is the promotion of energy service contracting in the HE sector. At present, take-up is very limited and there is widespread scepticism

Table 4.18 Recommendations for best practice in university energy management

Energy policy	This should establish principles, establish a structure of responsibility for energy management, assess risks, set objectives, and indicate measures to achieve those objectives. The policy should have the commitment of senior management and include quantitative targets for consumption or costs which are incorporated into the institutions' Estates Strategy.
Responsibility for energy management	EEO guidelines recommend a full-time energy manager for each £1m of utilities spend, while HEFCE recommend a higher ratio (HEFCE et al., 1996a, p. 21). Responsibility is best consolidated in the estates department and effective management requires staff with managerial, technical and 'marketing' skills. An energy committee should be established, integrated with the existing committee structure and providing both accountability of energy management activities and a forum for channelling proposals and suggestions.
Energy information system	This includes accurate meter reading and adoption of M&T. Analysis should include adjustment for weather, occupancy and other factors, and ultimately more sophisticated measures such as CUSUM (Croner, 2000). Consumption should be compared against disaggregated benchmarks and the results used to identify problems and prioritise individual buildings. Information should be regularly disseminated to top management, relevant committees, departmental management and staff in an effective and easy to assimilate manner.
Energy management profit centre	Initial funding could be based on historic utilities budgets together with a capital budget for energy efficiency investment (up to 10 per cent of annual utilities bill). Investment criteria should be established in consultation with the finance department, and based upon a clear rationale linked to the organisation's cost of capital. Savings from energy efficiency investment or contract negotiation should become a source of future funding and the centre should be able to access external funding for low-risk investments with a high rate of return.

Include incentive terms in staff contracts	Salary and promotion prospects of energy management staff could be partially linked to performance in reducing energy consumption and costs. Careful design is required here to ensure that performance can be measured effectively, that factors outside the staff's control do not adversely affect performance, and that an appropriate balance is obtained between risks and rewards.
User consultation	Consultation is relevant for both minor and major projects and is particularly important where user involvement is critical for the acceptability and performance of a technology (e.g. lighting controls).
Energy controls	The long-term aim should be BEMSs for all buildings controlled centrally by the estates department. Standardisation on a single system and development of specification documents for consultants are key elements.
Energy audits	This should begin with the buildings prioritised by the M&T system and should aim to identify both low-cost opportunities and longer-term options for energy efficiency improvement.
Purchasing procedures	This requires the rigorous adoption of whole-life costing, inclusion of energy and environmental considerations in purchasing guidance and the establishment of consultation procedures with energy management staff for purchases of energy intensive equipment.
Maintenance procedures	Routine maintenance work should be reviewed against an energy efficiency check list. Maintenance and refurbishment should be a primary source of efficiency improvement and should routinely include cost-effective efficiency measures.
Capital projects	Measures should include the inclusion of energy efficiency and minimisation of life-cycle costs as a key criterion of the project brief; development of energy efficiency guidance for architects and consultants; and the involvement of energy management staff in the preparation of design and engineering briefs and the evaluation of subsequent design and engineering documents.

of the value of energy service contracts. This is despite the general trend to contracting out non-core services in the UK public sector (Domberger and Jensen, 1997) and the more extensive take-up of such contracts in areas such as the health service. If government considers that ESCOs could save public money in the HE and other sectors, a case could be made for more extensive promotional effort. The objective would be to increase confidence in the energy services model and to reduce the transaction costs of contracting. Options for the latter include the development of model contracts that ensure institutional interests are protected, and the simplification of public procurement rules. Employment protection is another key issue, as staff and union opposition is one of the biggest obstacles to the expansion of energy services.

6.3 National Measures

All the above measures can be supported and encouraged by climate policy at the national and EU levels. Since the fieldwork was conducted there have been a series of important developments in UK and EU climate policy, the most relevant of which are the climate change levy (CCL), reform of the building regulations, market transformation programmes and the emergence of emissions trading (DETR, 2000a). The implications of these for the HE sector are briefly assessed below.

Climate change levy

The CCL, which came into force in April 2001, is a revenue neutral energy tax for business, commerce and the public sector which increases the unit price of electricity by 0.43p/kWh, and that of gas and coal by 0.15p/kWh.[56] Household energy use, oil use, CHP fuel and renewable electricity are all exempt. This corresponds to a 12 per cent increase in the average price of industrial electricity in 1999, and a 27 per cent increase in the price of gas (DTI, 2000).[57] The CCL is expected to raise £1 billion/year, but most of this is returned in the form of a 0.3 per cent cut in employers' national insurance contributions. The government chose a downstream tax, combined with indirect treatment of electricity[58] to avoid energy price increases for low income households, and chose an energy tax rather than a carbon tax to protect what remains of the UK coal industry. These decisions have undermined the economic efficiency of the scheme.

The shift in the tax burden from employment to energy should raise the profile of energy management and assist in addressing many of the identified barriers. While no-regrets energy opportunities are cost-effective at current energy prices, the key to capturing those opportunities lies in attracting management attention. The CCL does this in an effective way and should encourage improvements in energy management practices. As an illustration, the CCL should increase annual energy costs at university C by approximately 10 per cent, while the tax package overall should increase university C's expenditure by £30k/year. The exclusion of residential accommodation and some 'charitable' research activities, together with the high proportion of total costs accounted for by staff, suggests that many universities will be net winners from the CCL package.

Building regulations

The UK climate programme includes revisions to the building regulations which are expected to lead to a 25 to 30 per cent reduction in carbon emissions from new non-domestic buildings (DETR, 2000b). Important new features include standards on ventilation and air conditioning and an extension of the regulations to building refurbishment.[59]

Despite these amendments, UK regulations remain some distance behind other European countries and there is a substantial gap between regulated and best-practice performance.[60] While regulations provide an effective means of ensuring minimum standards, they fail to address the 'system' aspects of energy efficient design and are unlikely to encourage innovative green buildings (Gann et al., 1998). These points are explored further in Chapter 7, where it is argued that regulations need to be supplemented by wider changes in the organisation of the UK construction industry.

The EU has introduced a Directive on the energy performance of buildings, which includes the development of a standardised methodology for estimating carbon intensity (European Commission, 2002). This, or comparable methodologies, could provide a useful basis for performance targets for new and refurbished buildings in the HE sector.

Market transformation programmes

For many energy technologies (for example, computers), energy efficiency is likely to remain a marginal concern despite changes in organisational procedures. In these instances, both public and private benefits may be derived from transforming the market for such technologies. This may be through imposing mandatory standards or encouraging energy efficiency improvements through voluntary agreements with manufacturers. Standards overcome information barriers in energy service markets and reduce the risk to retailers of stocking energy efficient products.

Market transformation programmes are well established within the EU, but need expansion into the technologies appropriate for the non-domestic sector. The UK Market Transformation Programme has produced review papers of different technologies and sectors,[61] but further promotional measures require action at the EU level.

Emissions trading

Perhaps the most significant development in climate policy since the signing of the Kyoto Protocol has been the emergence of emissions trading schemes at the national, regional and global level. It now seems inevitable that emissions trading will provide a central component of EU and UK climate policy for years to come and will supplant many existing national regulations (Sorrell, 2003).

The EU is introducing a Europe-wide emissions trading scheme in 2005 which will place a cap on some 45 per cent of EU CO_2 emissions (European Commission, 2003). The scheme is initially targeted at large energy-intensive industries but will affect the university sector in two ways. First, through the involvement of electricity generators in the scheme, which could lead to substantial increases in electricity prices. And second, through the aggregate 20MW threshold for the inclusion of combustion plant in the scheme, which is sufficiently small to capture a number of campus universities – including university C. These universities will face a cap on their direct emissions of CO_2 which they must either meet through abatement measures or through purchasing additional allowances. If abatement costs are low, such universities may be able to profit from the sale of surplus allowances. This should be a very effective incentive to improve energy efficiency.

SUMMARY AND CONCLUSIONS

Higher education is representative of much of the public and commercial sector in that energy consumption is both substantial and increasing, but energy forms a small proportion of total costs and hence is widely ignored. The potential for cost-effective investment is significant, but a wide range of barriers prevent opportunities from being taken up. Barriers result from the transaction costs associated with identifying and implementing energy efficiency opportunities and the costs of overcoming information asymmetry and split incentive problems within large and complex organisations. Coupled with limitations on individual decision-making, this leads institutions to 'leave money on the floor'. While this neglect is particularly pronounced in the HE sector, both the wide variations in performance between the case study institutions and the existence of numerous best-practice examples suggest that it is not inevitable.

Hidden costs, particularly in the form of salary overheads, partly explain the severe time constraints on energy management staff, while transaction costs and agency relationships partly explain the rigid capital budgeting rules throughout the sector. But while these costs are hard to quantify, they do not fully explain the neglect of cost-effective opportunities. Institutions do not appear to have fully assessed the costs and benefits of energy management and many devote insufficient resources to this activity. Similarly, proven initiatives such as establishing an energy management cost centre or using energy service companies could overcome capital constraints, even in the context of severe financial difficulties. While transaction costs are partly inherent in the nature of energy efficiency, they may also be reduced through changes in organisational structures, routines and decision-making procedures.

As a consequence, there appears to be considerable scope for improving energy management practices and a wide range of best-practice examples to draw on. But these will need to be supported and encouraged by policy initiatives at the sector and national level. At present the reluctance of the HEFCs to intervene in energy management may be contributing to the poor performance of HE relative to other public sectors such as health. Benchmarking, reporting and targeting initiatives are possible and may stimulate improved value for money. And market-based instruments such as the CCL and emissions trading should help to place energy issues on the agenda of senior management.

Even with these changes, energy efficiency is likely to remain a marginal concern for most HE institutions. Here government policy may help in two key areas. First, market transformation initiatives may improve the energy efficiency of generic equipment such as computers and lighting; and second, reform of the construction industry may overcome barriers to energy efficiency in new build and refurbishment. This last point is the topic of Chapter 7.

NOTES

1. UK university case study interview, summer 1999.
2. All monetary values in this chapter are expressed in UK sterling (£), where £1 ~ €1.4.
3. There are separate funding councils for England (HEFCE), Wales (HEFCW) and Scotland (SHEFC), while Northern Ireland is the responsibility of the Department of Education for Northern Ireland.
4. One quarter of institutions needed to make major repairs to over half of their estate (HEFCE, 2001). Requirements for health and safety compliance, disabled access and expansion of space have created an additional drain on resources (HEFCE, 1998).
5. These are buildings of historical or architectural interest for which restrictions apply on how they can be internally or externally modified. In some circumstances, listed status may restrict the opportunities for improving energy efficiency.
6. The former polytechnics and some other higher education institutions were designated as universities in 1992, thereby ending the two-tier provision of degrees.
7. These were based on a combination of published government data and the results of energy monitoring at 25 sites.

8. Concerning heating, ventilation, air conditioning, controls, office equipment and building fabric.
9. The federal colleges of the University of London were treated as a single institution and the six institutions used as case studies were excluded. Some respondents were full-time energy managers, but more usually their job description suggested wider responsibilities, such as maintenance supervisor, mechanical engineer or procurement manager.
10. A non-valid reply means that no answer was given.
11. It seems reasonable to assume that a response to the questionnaire indicates an interest in, and willingness to devote time to, energy related issues. This, together with the small sample size, suggests that the results should be treated with caution.
12. See http://www.macc2.org.uk.
13. Monitoring and Targeting (M&T) is the UK term for a comprehensive energy data collection and analysis system. M&T is supported by a variety of software packages (Croner, 2000).
14. A wide range of sources from the UK Energy Efficiency Best Practice Programme were used to compile this list, including the 'Fuel Efficiency Booklet' series (for example, EEO, 1995).
15. Contrast, for example, the £50k/year energy efficiency budget at university C to the £4m/year spent on capital projects.
16. UK university case study interview, summer 1999.
17. Faced with a small budget and relatively little influence, the environmental adviser at university C concentrated on regulatory compliance (for example, waste) and low-cost initiatives that would stimulate staff interest. Energy efficiency was almost entirely ignored.
18. This is probably a high estimate, as a substantial proportion relates to expanding the system into new buildings, or extending to lighting control. More detailed figures on BEMS overhead costs were not obtained.
19. UK university case study interview, summer 1999.
20. UK university case study interview, summer 1999.
21. UK university case study interview, summer 1999.
22. The extent of these problems varied with individual circumstances. Campus universities with only one department per building had fewer physical obstacles, but faced substantial opposition from high profile science and engineering departments.
23. UK university case study interview, summer 1999.
24. UK university case study interview, summer 1999.
25. UK university case study interview, summer 1999.
26. HEFC guidance is via the Joint Procurement Policy and Strategy Group (JPPSG), formed in 1996. UK government guidance is developed in a range of publications (HM Treasury, 1998), while EU Directives on public sector procurement apply to purchases of >£160k for goods and services and £4m for capital projects.
27. There was a similar neglect of environmental criteria in purchasing procedures, even in those institutions with an established environmental policy. Consortia were beginning to require suppliers to be registered to an accredited EMS, but this gave no guarantee of the environmental performance of their products.
28. This finding is backed up by the results of a National Audit Office (NAO) investigation into university purchasing, which found both limited use of WLC and inadequate application of WLC techniques (NAO, 1999, p. 53).
29. Here, the computer department disliked energy saving software as there was a risk that people would turn computers off accidentally, since they look like they are not being used. Their priority was providing a service, not delivering energy cost savings.
30. 26mm tubes are not suitable for all existing fittings, but there were many instances at university C where they could have been used.
31. UK university case study interview, summer 1999.
32. UK university case study interview, summer 1999.
33. UK university case study interview, summer 1999.
34. UK university case study interview, summer 1999.

35. UK university case study interview, summer 1999.
36. UK university case study interview, summer 1999.
37. UK university case study interview, summer 1999.
38. This is for the best-practice system at university D. A large contribution derives from the five person days/month for meter reading, but electronic reading would require substantial capital investment.
39. This is itself a consequence of time and capital constraints. The circular argument is that investment is needed to obtain this information, but without this information it is difficult to know whether such investment would be worthwhile.
40. UK university case study interview, summer 1999.
41. For example: it was cheaper to use less efficient lamps with infrared movement sensors since the frequent switching made energy efficient lamps burn out more regularly; users often opposed lighting controls and requested override facilities; and photocell controls required considerable investment in commissioning to achieve the promised savings.
42. To the extent that survey respondents are biased towards the 'better' institutions, the quality of information systems in the sector may be worse than indicated here.
43. UK university case study interview, summer 1999.
44. This works well in the public sector because of the relative absence of competitiveness concerns, but it could be problematic in industrial sectors such as brewing (Chapter 5).
45. The University of Manchester is a notable and early example. See EEO (1994).
46. UK university case study interview, summer 1999.
47. UK university case study interview, summer 1999.
48. This approach had not been adopted by other case study institutions, with predictable consequences.
49. In 2002, the UK Energy Efficiency Best Practice Programme was taken over by the newly-formed Carbon Trust and renamed Action Energy. The Trust is making substantial changes to the style, content and delivery of this programme which should help to make it more effective.
50. Options include raising tuition fees, a graduate tax or giving universities an endowment to invest. This is a controversial topic, but the increased use of private sector finance seems inevitable.
51. The previous target set for government buildings was to reduce total energy consumption by 20 per cent over the period 1990–2000. In practice, only a 17 per cent reduction was achieved (DEFRA, 2001).
52. Carbon emissions from electricity consumption are estimated using a fixed emissions factor, but renewables are treated as zero carbon.
53. These are to: a) reduce the level of primary energy consumption by 15 per cent, or reduce aggregate carbon emissions by 0.15 MtC from 2000 to 2010; b) achieve an energy intensity of 35-55GJ/100m^3 in new buildings and major redevelopments/refurbishments; and c) achieve an energy intensity of 55-65GJ/100m^3 in all existing facilities.
54. Personal communication with NHS Estates and representatives of NHS Trusts.
55. Personal communication with HEFCE.
56. The fuel price increases correspond to £8.2/tCO_2 for gas, £4.6/tCO_2 for coal and £9.4/tCO_2 for electricity (using a fixed emission factor of 0.43 kgCO_2 per kWh of delivered electricity).
57. These are unit prices and exclude standing charges. The percentage change will be greater for large sites as unit prices are lower.
58. A direct treatment entails taxing the fuel input into electricity generation, while an indirect treatment entails taxing electricity at the point of consumption.
59. For refurbishment the requirements are less strict, less comprehensive, open to interpretation (for example, 'reasonable provision') and may vary with individual circumstances (DETR, 2000a, p. 101).
60. For example, 7.1kgC/m^2/year for a new naturally ventilated office compared to 2.3kgC/m^2/ year for best practice.
61. http://www.mtprog.com/.

REFERENCES

Ali Khan, S. (1996), *Environmental Responsibility: A Review of the 1993 Toyne Report*, Department of the Environment, Department for Education and Employment and Welsh Office, HMSO, London.

Ashley, A., M. Elsayed, J.F. Grant, M.D. Kelley, N.D. Mortimer and J.H.R. Rix (1997), 'Energy use in the non-domestic building stock: 1997 catalogue of results', Report No. SCP 4/9, Resources Research Unit, School of Urban and Regional Studies, Sheffield Hallam University, Sheffield.

Bruhns, H., P. Steadman and H. Herring (1997), *Activity Classification and the Structure of the Non-domestic Building Stock Database*, Centre for Configurational Studies, Open University, Milton Keynes.

Croner (2000), *Croner's Energy Management Guide*, Croner CCH Group Ltd, Kingston, Surrey.

Dearing, R. (1997), 'Higher education in the learning society', Report of the National Committee of Enquiry into Higher Education, HMSO, London.

Department of the Environment, Food and Rural Affairs (2001), *Greening Government Third Annual Report*, HMSO, London.

Department of Environment, Transport and the Regions (1998), *Non-domestic Building Energy Fact File*, Construction Research Communications, London.

Department of the Environment Transport and the Regions (2000a), *Climate Change: The UK Programme*, HMSO, London.

Department of the Environment, Transport and the Regions (2000b), 'The Building Act 1984: Building Regulations: proposals for amending the energy efficiency provisions', a Consultation Paper issued by Building Regulations Division, London, HMSO.

Domberger, S. and P. Jensen (1997), 'Contracting out by the public sector: theory, evidence, prospects', *Oxford Review of Economic Policy*, **13**(4), 67–77.

Department of Trade and Industry (2000), *Digest of United Kingdom Energy Statistics 2000*, HMSO, London.

Energy Efficiency Office (1993), 'Organisational aspects of energy management', General Information Report 12, ETSU, Harwell.

Energy Efficiency Office (1994), 'Energy management at Manchester University', Good Practice Case Study 150, ETSU, Harwell.

Energy Efficiency Office (1995), *Economic Use of Electricity in Buildings*, Fuel Efficiency Booklet No 9, ETSU, Harwell.

European Commission (2002), 'Directive 2002/91/EC on the energy performance of buildings', *Official Journal of the European Communities*, 4.3.2003, L. 1/65-71, Brussels.

European Commission (2003), 'Directive 2003/87/EC establishing a scheme for greenhouse gas emissions allowance trading within the Community and amending Council Directive 96/61/EC', *Official Journal of the European Communities*, L. 275/32-46, Brussels.

Fisher, J., W. Blyth, S. Collings, S. Boyle, J. Wilder, G. Henderson and M. Grubb (1998), 'Prospects for energy saving and reducing demand for energy in the UK', Paper prepared as background to the study on Energy and the Environment by the Royal Commission on Environmental Pollution, http://www.rcep.org.uk/studies/energy/98-6063/fisher.html.

Forum for the Future (1997), *Environmental Management Systems: A Guide for the HE Sector*, HE 21 Project, Forum for the Future, London.

Forum for the Future (1999) *Best Practice Bulletin: Purchasing*, HE21 project, http://www.he21.org.uk.

Gann, D., Y. Wang and R. Hawkins (1998), 'Do regulations encourage innovation? – the case of energy efficiency in housing', *Building Research and Information*, **26**(4), 280–96.

Grout, P.A. (1997), 'The economics of the private finance initiative', *Oxford Review of Economic Policy*, **13**(4), 53–66.

Higher Education Funding Council for England (1997), 'Consultation paper 1/97: the role of the Council in relation to Estates', HEFCE, Bristol.

Higher Education Funding Council for England (1998), 'Analysis of estate returns 1997', Report 98/04, HEFCE, Bristol.

Higher Education Funding Council for England (2000a), 'Analysis of 2000 financial forecasts and annual operating statements', Report 00/53, HEFCE, Bristol.

Higher Education Funding Council for England (2000b), 'Model financial memorandum between the HEFCE and institutions', Report 00/25, HEFCE, Bristol.

Higher Education Funding Council for England (2001), 'Estates Management Statistics Project: Annual Report 2001', Report 01/72, HEFCE, Bristol.

Higher Education Funding Council for England, Scottish Higher Education Funding Council, Higher Education Funding Council for Wales, Department of Education Northern Ireland (1996a), *Energy Management Study in the Higher Education Sector: National Report*, HEFCE, Bristol.

Higher Education Funding Council for England, Scottish Higher Education Funding Council, Higher Education Funding Council for Wales, Department of Education Northern Ireland (1996b), *Energy Management Study in the Higher Education Sector: Management Review Guide*, HEFCE, Bristol.

Higher Education Funding Council for England, Scottish Higher Education Funding Council, Higher Education Funding Council for Wales, Department of Education Northern Ireland (1996c), *Energy Management Study in the Higher Education Sector: Energy Management Benchmarking Software – User Guide*, HEFCE, Bristol.

Higher Education Statistics Agency (2000), *Higher Education Statistics for the United Kingdom 1998/99*, HESA Services Ltd, Cheltenham.

HM Treasury (1998), *Procurement Policy Guidelines*, HMSO, London.

Her Majesty's Government (1998) *Teaching and Higher Education Act of 1998*, HMSO, London.

Intergovernmental Panel on Climate Change (2001), *Working Group III Third Assessment Report*, Cambridge University Press, Cambridge.

Joint Procurement Policy and Strategy Group (1998), 'Whole life costing: a good practice guide for end users and all those involved in the procurement process', available at: http://www.jppsg.ac.uk/handbook/toolkit/wlc/.

Lovins, A.B. and L.H. Lovins (1997), *Climate: Making Sense and Making Money*, Rocky Mountain Institute, Colorado.

Mortimer, N.D., A. Ashley, C.A.C. Moody, J.H.R. Rix and S.A. Moss (1998), 'Carbon dioxide savings in the commercial building sector', *Energy Policy*, **26**(1), 615–24.

National Audit Office (1998), *The Management of Building Projects at English Higher Education Institutions*, HMSO, London.

National Audit Office (1999), *Procurement in the English Higher Education Sector*, HC 437 Session 1998-99, HMSO, London.

NHS Estates (2002), *New Environmental Strategy for the National Health Service*, HMSO, London.

Ramsden, B. (2001), *Patterns of Higher Education Institutions in the UK*, Universities UK, London.

Romm, J.J. (1999), *Cool Companies: How the Best Companies Boost Profits and Productivity by Cutting Greenhouse Gas Emissions*, Island Press, Washington DC.

Somervell, D. and T. Talbot (1991), *Educated Energy Management: Studies in the Management of Energy Resources in Educational Buildings*, University of Edinburgh, E & FN Spon, Edinburgh.

Sorrell, S. (2003), 'Back to the drawing board: implications of the EU Emissions Trading Scheme for UK climate policy', SPRU: Science and Technology Policy Research, University Of Sussex, Brighton, available from: http://www.sussex.ac.uk/spru/environment/research/drawingreport.pdf.

Sutherland, R.J. (1996), 'The economics of energy conservation policy', *Energy Policy*, **24**(4), 361–70.

Toyne, P. (1993), 'Environmental responsibility: an agenda for further and higher education', Report of a Committee on Environmental Education in Further and Higher Education, appointed by the Department for Education and the Welsh Office, HMSO, London.

5. Standing on a Burning Platform: Barriers to Energy Efficiency in the UK Brewing Industry

Steve Sorrell

'... currently we are on a burning platform. We must perform, otherwise we will be closed.'[1]

1. INTRODUCTION

Brewing is a process-based industry in which energy accounts for less than 5 per cent of production costs. As such, it is representative of much of the food and drink sector in that energy does not have the strategic importance we find in sectors such as iron and steel, but at the same time does not suffer the neglect that is found in light industry and the service sector. The bulk of energy is used in processes whose nature has changed little since Roman times, but which are now conducted on a large scale using efficient modern technology and sophisticated electronic controls. Breweries have competent engineering staff with in-depth knowledge of brewing processes and should have the skills to manage energy efficiently.

The UK brewing industry is also representative of much of UK manufacturing industry in that it is intensely competitive, with ownership becoming increasingly concentrated in a small number of large firms. The small, regional breweries that once formed the backbone of the industry are fighting for survival, and many have disappeared in the last few years.

This chapter explores the extent to which UK breweries are neglecting energy efficiency opportunities and the validity of competing explanations for this. The study is based on a postal survey of energy management at UK breweries, together with five in-depth case studies of individual firms. The chapter begins with an overview of the UK industry, including the nature of brewing processes and the pattern of energy use in breweries. The results of the postal survey are then reported, followed by a description and assessment of energy management practices at the case study breweries. These results are then combined to identify those barriers which appear of particular importance in the sector and to assess how they operate. This leads to brief recommendations on how the barriers may be overcome and how energy efficiency in the sector may be improved. Some key elements of UK climate policy are described and a critique is provided of the negotiated agreement on energy efficiency that the brewing sector signed with the UK government in 2001.

2. THE UK BREWING INDUSTRY

2.1 Beer and the UK Market

Beer is made through the fermentation of malted barley. Ales and stouts are fermented at room temperature and consumed a short time after production, while lager is fermented for longer periods at lower temperatures and requires conditioning by chilled storage for some weeks before consumption.

Ales have historically formed the main product of the UK industry, but the market is increasingly shifting towards lager. Indeed, from less than 1 per cent of the market in 1960, lager became the dominant drink in 1990 when it began to sell more than ale and stout combined. Brewery-conditioned ale is matured, filtered and pasteurised in the brewery and is served from aluminium kegs at the retail outlet, while cask-conditioned ale is not pasteurised and is served from casks containing yeast and conditioning agents. The majority of UK sales are for draught beer (casks and kegs), but in recent years there has been rapid growth in the market for bottled and canned beer. Casks and kegs are returned to the brewery for refilling, but less than a quarter of bottles are returnable. The market shifts from ales to lager, from draught to 'small pack' and from returnable to non-returnable bottles have important implications for brewery energy use.

Annual beer consumption in the UK has been declining at a modest rate since the 1970s and is now around 59 million hectolitres (hl). Total production in 1998 was 57 million hl and sales of beer amounted to £15 billion, or 1.7 per cent of UK GDP.

2.2 Industry Structure

Up to summer 2000, four companies accounted for around 84 per cent of UK beer production. The acquisition of the Whitbread and Bass brewing interests by the Belgian company, Interbrew, in June 2000 has increased concentration further. The industry also exhibits a high degree of vertical integration, with most companies being involved in production, wholesale and retail. A commonly used classification scheme is:

- *national brewers:* large brewers whose businesses extend across the UK and who own 'tied' pubs selling the brewers' own brands;
- *brewers without tied estates (BWTEs):* large brewers who do not own pubs and who brew beer for sale to other brewers or independent wholesalers;
- *regional brewers:* whose business has historically been focused on a single region; and
- *micro-brewers:* who brew very small volumes and often sell to a single retail outlet.

The distinctions between these categories are now less clear-cut due to market changes and government intervention to increase competition.

Prior to June 2000, there were three national brewers, three BWTEs, 60 regional brewers and around 200 micro-brewers. Following the Interbrew takeover, there are now only two national brewers (Interbrew and Scottish Courage) accounting for around 63 per cent of the UK market. Carlsberg-Tetley (a BWTE) is usually grouped with the national brewers and the three (previously four) are referred to collectively as the 'major brewers'.

Excluding micro-brewers, the total number of breweries has declined steadily over the last 20 years, from 142 in 1980 to 94 in 1998. Closures and concentration of production have occurred in both the majors and regional companies, encouraged by a series of mergers and acquisitions. As Table 5.1 shows, beer production is increasingly dominated by a small number of large breweries. Twenty-five breweries producing more than 0.5 million hl/year accounted for 94 per cent of recorded production in 1998, while 17 sites producing more than 1 million hl/year accounted for over 80 per cent.

Table 5.1 Size distribution of UK breweries in the 1998 BLRA energy survey

Output range (th hl/year)	No. responding	Total production (th hl/year)	% recorded production
<100	15	578	1.1
100-499	13	2717	5.2
>500	25	48824	93.7
Total	53	52119	100.0

Note: The survey received responses from 53 breweries, representing 92 per cent of UK production.

Source: Brewers and Licensed Retailers Association Energy Survey 1998.

Declining domestic demand has led to overcapacity, increased competition, reduced margins and an emphasis on exports and product differentiation. The wholesale price of beer has fallen, despite increases in retail prices. The government has reduced brewers' control over retail outlets, leading to a shift in market power from brewers to retailers. As a result, both the majors and the regional companies have been aggressively developing their business in hotels, pub chains, restaurants and other parts of the leisure industry.[2] This all makes for difficult times in the brewing industry, with repeated cost-cutting drives, staff reductions and considerable uncertainty over future closures.

2.3 The Brewing Process

The production of beer can be divided into four stages: brewing, fermentation, processing and packaging (Figure 5.1). The processes are traditional, with limited technical innovation, although the larger breweries have invested heavily in electronic controls. Conservatism and tradition are particular features of the smaller breweries producing cask ale.

Malt is milled at the brewery to extract the key constituents and separate the husk. Hot water is then added to the malt to form a mash, with the mixture held in a vessel termed the mash tun. The enzymes in the malt are reactivated and the mash is left to stand while the starch is converted into sugars. The extract, termed the wort, is filtered through the bed of grains leaving behind the husks.

Hops are then added to the wort and the mixture boiled in a vessel known as the copper. The boiling process turns the wort bitter, inactivates the remaining enzymes and extracts oils and resins from the hops which add flavour and help to preserve the final product. The wort is then filtered to remove the spent hops and excess protein, and cooled before transferring to the fermenting vessel. Fermentation is initiated by the addition of yeast which converts the sugars to alcohol.

When fermentation is complete the beer is cooled. Cask-conditioned beer is transferred to metal or wooden casks and clarified by the addition of purified protein which sediments the yeast. Most beer is conditioned within the brewery through filtration, pasteurisation and, in the case of lager, carbonation. The beer may either be pasteurised in bulk and packaged in pre-sterilised kegs, or packaged in bottles and cans and pasteurised in the sealed container.

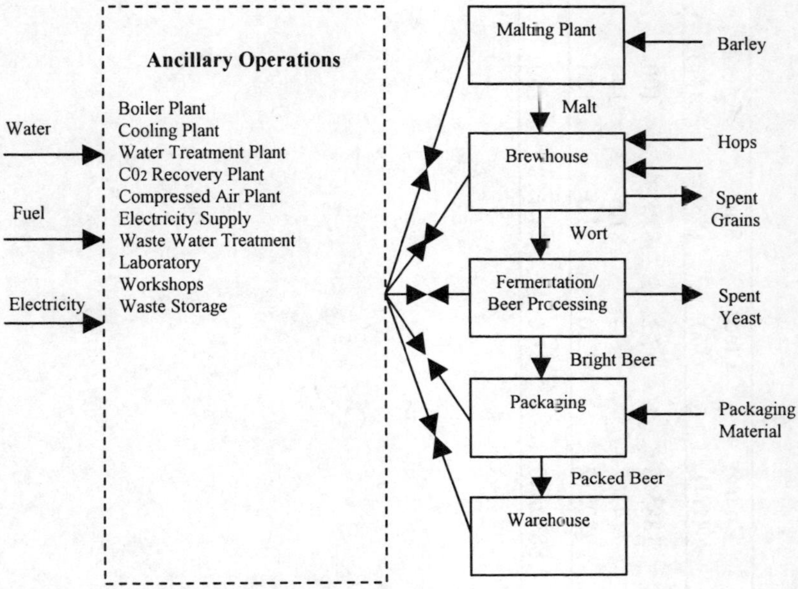

Source: United Nations Environment Programme (1996), p. 19.

Figure 5.1 Schematic layout of a brewery, showing resource streams

2.4 Energy use in the Brewing Industry

Table 5.2 summarises energy use by size category in the UK brewing industry in 1998. The 53 breweries responding to the survey used a total of 1919TJ of electricity and 6963TJ of fuel, corresponding to 0.53 per cent of industrial electricity use and 0.58 per cent of industrial fuel use. In total, the recorded breweries accounted for 0.13 per cent of UK delivered energy use.

In 1989, energy accounted for 3.97 per cent of production costs in the combined brewing and malting sector (DTI, 1994). More recent data is not available, but a combination of rationalisation of production at the most efficient sites, improved energy efficiency and substantial falls in energy prices should have reduced this figure. In 1998, the total expenditure on energy by the sector was approximately £35 million.

Some 22 per cent of brewery energy requirements are supplied by electricity, with the proportion being higher in large breweries (Table 5.2).[3] The difference may be partly accounted for by the higher proportion of cask ale produced in small breweries, and the smaller volume of bottle and can production. The 25 large breweries account for 92 per cent of electricity use and 89 per cent of fuel use.

Table 5.2 Energy use by size category in the UK brewing industry 1998

Capacity (th hl/year)	Sample size	Elec. (GJ)	Fuel (GJ)	Energy (GJ)	Elec. (%)	Fuel (%)	Energy (%)
<100	15	32367	156194	188562	17.2	82.8	100
100–499	13	131519	582743	714262	18.4	81.6	100
>500	25	1755092	6224924	7980017	22.0	78.0	100
Total	**53**	**1918979**	**6963862**	**8882840**	**21.6**	**78.4**	**100**

Source: Brewers and Licensed Retailers Association (1999).

As a result of biennial surveys by the Brewers and Licensed Retailers Association (BLRA),[4] the UK brewing industry has good data on trends in specific energy consumption (SEC), measured in MJ per litre of beer (BLRA, 1999). This shows that the average SEC of the sector has decreased from 3.03MJ/l in 1976 to 1.70 MJ/l in 1998, a reduction of 44 per cent. There was a fairly continuous improvement in energy efficiency up to 1992, little progress during the 1990s, and then a jump to 1.70MJ/hl in the 1998 survey. Energy consumption is correlated with water consumption, which declined by 31 per cent over the same period.

Table 5.3 shows specific energy and water consumption by brewery capacity. This demonstrates that larger breweries are more energy efficient. On average, small breweries use twice as much energy to produce a litre of beer than do large breweries. Table 5.3 also illustrates the wide range of specific energy consumption within each brewery size range. Thus, the least efficient small breweries use six times as much energy per litre of beer than the most efficient. More detailed analysis would be required to identify the reasons for this, but the results suggest that performance in energy management varies widely throughout the sector.

Table 5.3 Specific energy and water efficiency of UK breweries in 1998 by size range

Capacity (th hl/year)	**SEC (MJ/l)**	**SEC range (MJ/l)**	**SWC (hl/hl)**	**SWC range (hl/hl)**
<100	3.26	1.20–7.17	7.78	3.06–19.43
100–499	2.63	1.09–4.51	8.12	3.19–14.17
>500	1.63	0.96–2.32	5.90	3.51–9.94
Mean	1.70		6.04	

Source: Brewers and Licensed Retailers Association (1999).

The SEC of an individual brewery is strongly influenced by the shift pattern, product mix and packaging mix. Electricity use is greater for lager production than for ale, due to greater refrigeration needs, while energy use for packaging varies from around 3 per cent of the total for bulk tanks to 56 per cent for returnable bottles (EEO, 1992). The BLRA has developed formulae to allow these differences to be taken into account, and has defined 'best practice' figures for different parts of the brewing process (EEO, 1992; BLRA, 1998). In 1996, the average energy efficiency of the industry was 16.5 per cent below best practice, with the very largest breweries 11 per cent below, and breweries producing <200th hl/year, 50 per cent below. The spread of efficiencies within each size category is very large, but particularly so for small breweries

There is little data available on the breakdown of energy use by end-use. This will vary between breweries and also within the same brewery depending on the current packaging mix. Table 5.4 shows some estimates for a brewery producing keg ale. Steam use is dominated by mashing and wort boiling, while electricity use is dominated by refrigeration. If the brewery were producing returnable bottles, the steam use for packaging would exceed all other uses combined.

Table 5.4 Energy consumption by end-use in a kegging brewery

Type	Area	%
Thermal energy	Brewhouse	20–50
	Packaging	25–30
	Space heating	<10
	Utilities	15–20
Electrical energy	Refrigeration	30–40
	Packaging	15–35
	Compressed air	10
	Brewhouse	5–10
	Boiler house	5
	Lighting	6
	Other	10–30

Source: BECOM (1995).

Heat recovery is common in breweries and most UK breweries use heat from the cooling wort to preheat brewhouse water ('liquor') (Harris, 1979). A less common technique is to condense the water evaporated during wort boiling (10 per cent by mass) and to use the heat for preheating liquor, keg washing or other tasks. The amount of heat that can be recovered depends on the brewing schedule and the manner in which the water storage system is used. A common obstacle to additional heat recovery is an inability to find a use for the hot water produced.

In terms of energy management, the small, independent regional breweries are very different organisations from the large sites owned by the majors. In small breweries, energy is one of many responsibilities of the chief engineer, supported by a small number of technical staff. In contrast, large breweries commonly have a dedicated energy/utilities manager, whose primary responsibility is the control of a site 'service centre', including water treatment, heat and electricity production, compressed air, industrial gases and refrigeration. In both cases, attention is primarily focused on process technology, rather than buildings energy use, as this accounts for the greatest proportion of consumption. Brewing and engineering staff are separate in the large breweries, and frequently have different perspectives and priorities.

2.5 Energy Saving Potential

In the UK, the most authoritative source on industrial energy efficiency potential is the 'GAD' database and associated Energy Use Simulation Model (ENUSIM), developed by ETSU (now Future Energy Solutions, or FES) at AEA Technology (AEAT) (ETSU, 1996).[5] This database exploits the knowledge built up by ETSU through their management of the government Energy Efficiency Best Practice Programme (EEBPP).[6]

In 2001, ETSU published energy projections for the combined brewing, cider making and malting sector (ETSU, 2001).[7] These gave a 'business as usual' projection of a 5 per cent increase in SEC[8] between 2000 and 2010, while adoption of 'all cost effective' (ACE) energy efficiency measures would achieve a 14.5 per cent improvement. These projections assumed normal plant replacement schedules, incorporated assumptions on replacement, retrofit and improved energy management, and assessed cost-effectiveness by examining *all* potential savings, not just energy.

ETSU defined cost-effectiveness in the food and drink sector as the use of two year paybacks for retrofit and two to four years for replacement (ETSU, 1999). While this may represent current industry practice, it implies discount rates that substantially exceed the cost of capital to the sector. To use short paybacks as a basis for assessing cost-effective potential is to effectively assume that stringent investment criteria are rational given problems of business risk, hidden costs and other factors. But ETSU also claimed that the ACE scenario was optimistic as it assumed no restriction on capital availability or management time. Since the latter is one measure of the 'hidden' overhead costs of energy management, the logic of choosing such short paybacks in the first place becomes even more questionable.

Another area of concern relates to the high level of sectoral and technology aggregation in the GAD model – the options included represent only a subset of the actual opportunities available (Sorrell and Smith, 1999). In sum, while the projections suggest relatively small opportunities to 'cost-effectively' improve energy efficiency, the definition of cost-effectiveness is questionable and the range of opportunities considered may be inadequate.

Table 5.5 Coverage of the 1999 BLRA utilities and environmental survey

Category	Production (th hl/year)	Survey population	Total production of group (th hl)	No. of responses	Specific energy consumption (MJ/l)	Response as % of population
Small	< 100	28	578	15	3.26	53
Medium	100–500	20	2717	13	2.63	65
Large	> 500	27	48824	25	1.63	92
Total		**75**	**52119**	**53**	**1.70**	**71**

Table 5.6 Coverage of the 1999 SPRU brewery energy survey

Category	Production (th hl/year)	Survey population	Total production of group (th hl)	No. of responses	% of responses in each category	Response as % of population
Small	< 100	28	578	13	32.5	46
Medium	100–500	20	2717	12	30.0	60
Large	> 500	27	48824	13	32.5	48
Total		**75**	**52119**	**40**	**100**	**53**

3. UK BREWERY ENERGY MANAGEMENT SURVEY

A postal survey was carried out on energy management practices, technology adoption and perceived barriers to energy efficiency in UK breweries in 1999. This survey was distributed alongside the April 1999 BLRA Utilities and Environmental survey (Table 5.5), which is used to calculate SEC figures (BLRA, 1999).[9] A total of 40 replies were obtained, representing 53 per cent of the survey population and 77 per cent of the respondents to the BLRA survey (Table 5.6). Responses were split fairly equally between small, medium and large breweries. The results are summarised below in terms of the percentage of valid replies to each question.[10] The small sample size must be borne in mind when interpreting results expressed as percentages, while the fact that the respondents were self-selecting is likely to have biased the results towards the more proactive institutions. Nevertheless, the results provide a useful picture of energy management practices in one half of UK breweries in 1999.

3.1 Self-assessment Matrix

Respondents were asked to complete the self-assessment matrix on energy management reproduced in Annex A (section 6). This matrix, which is well established in the UK, grades organisational performance from 0 (poor) to 4 (excellent) in six categories (EEO, 1993). Table 5.7 summarises respondents' performance in terms of overall weighted average scores. The overall average score was 1.9, which is reasonable but leaves considerable scope for improvement. The average score in each category corresponds to:

1. *Policy*: unadopted energy policy;
2. *Organisation*: energy manager, but with authority unclear;
3. *Communication*: contact with users through an ad hoc committee;
4. *Information*: monitoring and targeting reports based on supply meter data;
5. *Awareness*: some ad hoc staff awareness training;
6. *Investment*: same payback criteria as for other investments.

The results suggest that breweries are relatively good on information systems and investment, but poorer on organisation and communication. This is consistent with a technical approach to energy management that emphasises process control but gives lower priority to staff awareness and housekeeping. Performance improves with the size of the brewery, with a difference of 1.0 in the overall weighted average scores for small and large breweries.

Table 5.7 Self-assessment matrix: overall weighted average scores

Description	Small	Medium	Large	Total
Policy	1.00	1.36	2.73	1.74
Organisation	1.11	1.64	2.09	1.65
Communication	1.00	1.82	2.09	1.68
Information systems	1.56	1.82	2.64	2.03
Awareness	1.44	1.64	2.09	1.74
Investment	2.33	2.73	3.00	2.71
Overall	**1.41**	**1.84**	**2.44**	**1.93**

Note: Score from 0 (poor) to 4 (excellent). See Annex A (section 6) for details.

3.2 Energy and Environmental Management

Only nine respondents (29 per cent) had a formal energy policy, of which six were in the large category. A larger proportion (39 per cent) had adopted some form of environmental policy, including three-quarters of large breweries. No respondents were signatories to the government's *Making a Corporate Commitment* (MACC), which requires signatories to publish a formal energy policy and set targets for energy use.[11] MACC may be indicative of a stronger commitment to energy efficiency than in-house policies.

Over half (56 per cent) of respondents claimed to have conducted energy/environmental audits, including 85 per cent of large breweries and 46 per cent of small. Environmental management systems had been adopted by 28 per cent of respondents, but none were accredited to a recognised standard, such as ISO14001.

Table 5.8 Energy/environmental management (per cent of valid replies)

Description	Small	Medium	Large	Total
Formal energy policy	11	18	55	29
Signatory to MACC	0	0	0	0
Have conducted energy/environmental audits	46	30	85	56
Have environmental policy	15	20	77	39
Have environmental management system (EMS)	8	10	62	28
EMS is accredited	0	0	0	0

3.3 Energy Information Systems

A comprehensive data collection and analysis system for energy consumption is a prerequisite for effective energy management. Of particular importance is the level at which energy is metered and the frequency with which it is recorded.

Survey results showed substantial differences between large and small sites (Tables 5.9 and 5.10). Over three-quarters of small breweries metered electricity at the site level, compared to only a quarter of large breweries. Similarly, half of the large breweries metered heat and electricity at the process level, compared to none of the small breweries. Energy consumption was recorded daily by 11 per cent of respondents, with 51 per cent recording weekly, 27 per cent monthly and 11 per cent annually. More frequent recording can be effective in allowing faults and other problems to be rapidly identified and dealt with.

Table 5.9 Level at which energy use is metered (per cent of valid replies)

Energy	Level	Small	Medium	Large	Total
Electricity	Site	77	46	25	50
	Building	23	36	25	28
	Process	0	18	50	22
Heat	Site	78	55	27	52
	Building	22	27	18	23
	Process	0	18	55	26

Table 5.10 Frequency that energy use is recorded (per cent of valid replies)

Frequency	Small	Medium	Large	Total
Annually	31	0	0	11
Monthly	39	33	8	27
Weekly	31	58	67	51
Daily	0	8	25	11

Table 5.11 indicates how energy data is analysed and used. Ninety per cent of respondents monitored trends in consumption, 68 per cent adjusted for production output and 47 per cent used a Monitoring and Targeting (M&T) scheme.[12] Again, there were substantial differences between large and small breweries: for example, 92 per cent of large breweries used an M&T scheme, compared to only 15 per cent of small. The BLRA benchmarks were used by 58 per cent of respondents, including three-quarters of large breweries but

only 46 per cent of small. Given the 20 year history of these benchmarks and the prominence given to them by the trade association, this seems relatively low.

Table 5.11 Energy information systems and energy management (per cent of valid replies)

Description	Small	Medium	Large	Total
Monitor trends in energy use	77	92	100	**90**
Adjust consumption records for production output	54	58	92	**68**
Use a monitoring and targeting scheme	15	33	92	**47**
Compare consumption with sector benchmarks	46	50	77	**58**
Monitor savings from efficiency investments	50	55	62	**56**
Devolve responsibility for energy to cost centres	31	33	39	**34**

3.4 Investment in Energy Efficiency

Only 38 per cent of respondents gave a quantitative response to a question on investment criteria for energy efficiency. These reported a mean payback of 2.8 years, with little difference between large and small sites (Table 5.12). Four breweries (three large, one medium) reported using internal rate of return (IRR) or net present value (NPV), while five reported using simple payback, without stating the value. Overall, the results suggest that investment appraisal is unsophisticated and that short payback criteria dominate.

Participants were asked how much they agreed that 'a wide range of energy efficiency measures could be implemented that would yield paybacks of <4 years at current energy prices'. The results (Table 5.13) show that 48 per cent of respondents agreed or strongly agreed with this statement, with 19 per cent neutral and 33 per cent disagreeing. The suggested payback was long by sector standards, so it is interesting that a full one-third disagreed. Scepticism was more pronounced in the medium and small breweries, despite the fact that these have poorer SECs and are considered by analysts to have greater energy efficiency opportunities (EEO, 1992).

Table 5.12 Investment criteria normally applied for investments in energy efficiency

Criteria	No. of replies
Payback < 2 years	4
Payback < 3 years	8
Payback < 4 years	3
Payback < 5 years[a]	1
Payback – unspecified period	5
15 per cent IRR	1
Unspecified IRR	2
Other[b]	10

Notes:
a) Only if there are other environmental benefits; normally 3 years.
b) Responses included 'ad hoc', 'reasonable payback' and 'none – usually a by-product of investments for other purposes'.

Table 5.13 Wide range of energy efficiency measures with <4 year payback? (per cent of valid replies)

Description	Small	Medium	Large	Total
Strongly agree	8	0	8	6
Agree	50	18	54	42
Neutral	8	36	15	19
Disagree	25	18	23	22
Strongly disagree	0	18	0	6
Don't know	8	9	0	6

Respondents were fairly evenly split on whether energy efficiency improvements were largely a by-product of other investments or whether they derived mostly from investments whose primary purpose was to improve energy efficiency. A small majority (56 per cent) favoured the latter.[13]

3.5 Technology Adoption

Participants were given a list of 38 widely promoted energy efficient technologies and asked which they had adopted. The cost effectiveness of each technology will depend upon the specific circumstances of the brewery and whether replacement or retrofit is being considered. However, the majority of the listed technologies are claimed to give paybacks of less than five years (EEO, 1992).

Table 5.14 indicates the percentage of breweries which had adopted each technology. Widely adopted technologies included insulation of process plant, condensate recovery, power factor correction and variable speed drives. Technologies with relatively limited adoption included CHP, boiler economisers, high frequency fluorescents and various forms of heat recovery, including copper vapour heat recovery (CVHR). Some of these technologies may be inappropriate – for example, the boilers may be too small for economisers or the shift pattern may be inappropriate for CHP. However, the results suggest that many apparently cost-effective technologies continue to be overlooked.

Table 5.14 Adoption of energy efficient technologies/techniques at UK breweries

Area	**Technology/technique**	**Per cent adopted**
Brewing processes	Insulation of process plant and storage vessels	95
	Additional heat recovery from wort cooling	74
	Brewing at high specific gravity	45
	Reduction in wort boiling time	37
	Copper vapour heat recovery	21
	Steam eductor recompression of wort vapour	0
	Mechanical vapour recompression of wort vapour	0
Packaging	Variation of washing cycle with size of cask/keg	53
	Cask/keg line heat recovery	45
	Bottle washer/pasteuriser heat recovery	26
	Increase in beer racking temperature	18
Refrigeration	Improved systems for temperature control	76
	Placing maximum cooling duties at night	24
	Refrigeration heat recovery	11
	Coolant storage via ice tank or hydrated salts	8
Boilers/steam	Insulation of pipes, valves and flanges	92
	Condensate recovery	87
	Boiler blowdown heat recovery	45
	Use of boiler sequencing controls	42

	Recovery of flash steam	42
	Oxygen trim controls	37
	Installation of economiser	21
	CHP	5
Lighting	Replacement of mercury discharge lamps with fluorescents or high pressure sodium	61
	Replacement of 38mm fluorescents with 26mm?	56
	Replacement of tungsten filament lamps with compact fluorescents	53
	Use of photocell, acoustic or movement sensors?	47
	Use of high frequency electronic ballasts?	25
Electrical	Power factor correction	81
	Use of variable speed drives (VSD) in pumps, fans etc.	78
	Automatic switch-off of fans and pumps	56
	Specification of high efficiency motors on motor replacement	47
Compressed air	Regular inspection and repair of leaks	69
	Generation of air only when needed and at minimum required pressure	53
	Consideration of energy consumption when purchasing new compressors	50
	Pre-cool inlet air/duct from outside	33
	Compressor heat recovery	6
Overall mean		***43***

On average, the listed technologies had been adopted by 30 per cent of small breweries, 42 per cent of medium and 56 per cent of large. The fact that some technologies may be inappropriate for small sites only explains a portion of this difference. For example, nearly half of small breweries had no power factor correction, despite this being a cost-effective technology applicable in all instances.

3.6 Barriers to Energy Efficiency

Participants were asked to rate the importance of 17 suggested barriers to energy efficiency using a scale of often important (2), sometimes important

(1) or rarely important (0). As with the postal survey of UK universities, each of the 21 barriers was phrased in terms that would be meaningful to the survey participants, but most related to one of the six more general barriers listed earlier in Table 2.5. The results are summarised in Table 5.15, which shows the percentage of valid replies in each category, together with the weighted average score.

The overall mean score was 0.97, with the score for individual barriers ranging from 0.47 to 1.31. This suggests that a wide range of barriers are considered relevant, with few standing out as either very important or rarely important. 'Technology inappropriate at this site' was considered the most important barrier, closely followed by 'other priorities for capital investment' and 'lack of time'. Barriers considered of limited importance included 'lack of skills' and 'cost of staff replacement, retirement and retraining'. Perceptions of the importance of barriers were higher for small breweries than for large, particularly with regard to technical skills, capital constraints and staff awareness.

These results provide a valuable supplement to case studies and will be referred to subsequently.

Table 5.15 Perceptions on barriers to energy efficiency in UK breweries (per cent of valid replies)

Proposed barrier	**Often important**	**Sometimes important**	**Rarely important**	***Weighted score***
Technology inappropriate at this site	47	36	17	***1.31***
Other priorities for capital investment	38	54	8	***1.30***
Lack of time/other priorities	31	58	11	***1.20***
Technical risk	31	50	19	***1.11***
Energy objectives not integrated into operating, maintenance or purchasing procedures	38	35	27	***1.11***
Conflicts of interest within the organisation	39	33	28	***1.11***
Department/individuals not accountable for energy costs	36	36	28	***1.08***
Lack of staff awareness	19	61	19	***1.00***

Lack of information/poor quality information on energy efficiency opportunities	28	39	33	***0.95***
Loss of benefit (e.g. unreliability)	29	37	34	***0.94***
Lack of capital	31	28	42	***0.89***
Difficulty/cost of obtaining information on the energy consumption of purchased equipment	22	39	39	***0.83***
Cost of identifying opportunities, analysing cost effectiveness and tendering	16	49	35	***0.81***
Cost of production disruptions /hassle /inconvenience	22	33	44	***0.78***
Business/market uncertainty	22	33	44	***0.78***
Lack of technical skills	14	47	39	***0.75***
Cost of staff replacement, retirement, retraining	8	31	61	***0.47***
Overall mean	***28***	***41***	***31***	***0.97***

Note: Score 2 for often important, 1 for sometimes important and 0 for rarely important.

4. CASE STUDIES OF ENERGY MANAGEMENT IN UK BREWERIES

Case studies were undertaken of energy management practices within five breweries. These included both large and small sites and a mix of regional companies and brewery majors. In each case, up to four people, including the energy manager or equivalent, were interviewed and supplementary information was obtained from the brewery's responses to the BLRA and SPRU surveys. The interviews were semi-structured and used detailed protocols based on the theoretical framework developed in Chapter 2 (see Annex B).

Table 5.16 Summary of UK brewery case studies

Feature	Brewery A	Brewery B	Brewery C	Brewery D	Brewery E
Ownership	major	major	regional	regional	regional
Size category	large	large	medium	medium	small
Shift pattern	continuous	continuous	single	single	single
Production (th hl/year)	1987	1978	373	182	58
Packaging (th hl/year)	n/a	1978	604	234	58
% casks	0	0	47	41	97
% kegs	70	46	51	33	0
% small pack	30	54	3	25	3
Energy spend (£k/year)	800	1350	375	276	34
Utilities spend (£k/year)[a]	2300	2490	535	496	52
SEC (MJ/l)	1.73	1.41	< 2.6	2.7	n/a
Self-assessment score[b]	1.6	3.2	2.5	2.0	1.0
Energy policy	No	No	No	No	No
Environmental policy	Yes	Yes	Yes	No	Yes
ISO14001	working towards	considering	working towards	No	No
Meter reading frequency	weekly (manual)	hourly (electronic)	weekly (manual)	daily (manual)	annually (utility)
Meter reading level	cost centre	process	cost centre	site	Site
% of listed technologies adopted	35	76	60	55	<30

Use of targets	Yes	Yes	Yes	No	No
Comparison with benchmarks	Yes	Yes	Yes	No	No
Energy audits	No	Yes	No	No	No
Payback criteria	< 3 years	2 years	< 2 years	2–3 years	'when necessary'
Budget for energy efficiency investment	No	No	No	No	No

Notes:

a) Utilities here means energy and water costs together with effluent charges. Breweries C, E and B used their own bore hole, while A and D purchased mains water. Effluent charges were substantial for all breweries, and in the case of Brewery A exceeded total energy costs.

b) The average self-assessment score from the postal survey was 1.9.

Table 5.16 summarises the main features of each case study brewery. The largest (A) produces 57 times as much beer annually as the smallest (E). These are very different types of organisation and can be expected to differ in their energy efficiency performance, approach to energy management and the relevance of different types of barrier.

The following summarises energy management practices at the case study breweries.

4.1 Organisation

Only one site (B) had a dedicated energy manager. At the other sites, energy management was one of a number of responsibilities of the chief engineer or utilities engineer. The total staff time devoted to energy management was difficult to assess, but appeared to be below best-practice guidelines for at least three of the sites.[14]

Each brewery had competent engineering staff and much work was conducted in-house. In no case was there a dedicated budget for energy efficiency investment and interviewees emphasised that the majority of projects had objectives other than reducing energy use. Production was typically organised into cost centres, with engineering staff providing a service to each. The importance attached to energy efficiency varied between the sites, but in all cases the primary concern was process reliability and product quality, with energy efficiency being very much secondary.

A dominant theme was the heavy workloads and increasing time constraints on engineering staff. Companies had responded to the difficult business situation by using redundancies to cut costs, and this had only been partly compensated by increased automation.

4.2 Energy and Environmental Policy

None of the case study breweries had a formal energy policy, although brewery C included energy objectives within a broader environmental policy. Generally, there was scepticism of the value of policy statements and a view that energy efficiency should result from 'good engineering'.

Three of the sites had some form of environmental policy, while brewery E had won an environmental business award. The larger sites were seriously examining environmental management systems with the aim of accrediting to ISO14001. This was partly a result of supply chain pressure from large customers, coupled with broader considerations regarding investor attractiveness and public profile. For the smaller companies, good environmental performance fitted with the marketing image of a traditional business producing a natural product. But the impact of these initiatives on energy efficiency was marginal at best. Of far greater importance was the anticipated introduction of climate policy instruments, such as the Climate Change Levy (section 6.2 below).

4.3 Energy Information Systems

Energy information systems varied greatly depending upon the size of the site. Brewery B, had a highly sophisticated M&T system including electronic reading of sub-meters at hourly intervals and the use of exception reports to identify problems. While this system achieved substantial savings, installation costs were high (£2–3k per meter) which partly explains the continued use of manual metering at the other sites. At the other extreme, brewery E relied solely on utility bills and did not track energy consumption. Three of the sites compared performance against benchmarks, while four used monitoring data to set targets for cost centres.

While there was clearly scope for improving information systems, there were also large overheads, including the salary costs of engineering staff who devoted much time to running the systems. Nevertheless, there appeared to be a correlation between energy efficiency performance and the quality of the information system, which was independent of site size.[15]

4.4 Accountability and Incentives

Three of the sites charged cost centres for utilities consumption, but interviewees were sceptical of the value of such charging. The primary incentive was to stay within budget instead of identifying possible savings, and most staff could do little to influence energy use beyond housekeeping. Also, other priorities such as meeting production targets tended to dominate. As a result, the biggest improvements were believed to result from engineering activities such as the installation of new equipment.

4.5 Capital Budgeting and Investment

None of the breweries had a dedicated budget for energy efficiency investment and the energy manager or equivalent had to make a case for the funding of specific projects on their merits. The priority for capital investment was business development projects such as expanding the retail chain,[16] improving product quality and increasing capacity to accommodate transfer of production from other sites. Following this were business maintenance projects, such as essential replacement and hygiene. Pure energy saving investments had a low priority and were rare. Energy efficiency improvements were largely achieved as a by-product of investments undertaken for other reasons.

Two breweries used 3 year paybacks, two used 2 years, while brewery E invested 'when necessary', without calculating paybacks. Investment criteria were identical for all types of cost saving investment. There was no attempt to calculate the IRR or NPV, and the rationale for stringent paybacks was rarely explicit. The most common reason given was the short time horizons resulting from the risk of brewery closures, takeovers and mergers.

4.6 Purchasing and Policy Integration

At smaller sites, utilities staff were intimately involved in all new projects, so attention to energy efficiency depended largely on their knowledge and awareness. At the larger sites, project managers were separate from utilities staff and there was evidence of lack of consultation between the two groups:

> 'We tend to find stuff after it happens. A couple of weeks ago I had a discussion with the people over in small pack because they've put in a new cleaning system...it was very inefficiently installed, because the condensate was going straight to drain...by it going down the drain we're losing the water – we're paying to bring the water out of the ground, we're paying to put it back down the drain, and we're losing all the heat in it. People doing their own thing without involving our department costs us money.'[17]

Whole-life costing was frequently neglected in favour of meeting capital budgets, and utilities equipment could be oversized. In the absence of formal procedures for considering energy efficiency, consultation was ad hoc.

4.7 Awareness and Culture

The level of energy awareness among employees was uniformly reported to be poor. All sites had used a variety of awareness campaigns – such as posters and information stands – but all reported that these were ineffective. Awareness raising could potentially be assisted through accountability for energy costs, using the environmental impact of energy use to motivate staff, and moves towards 'team-working' and greater responsibility for cost centre performance. But all sites lacked a high-level champion for energy efficiency and all reported that awareness of energy efficiency was low within the 'brewing culture'. Tension between brewers and engineers was common:

> 'In theory I should be able to talk to the brewer and sort it out. But he is not interested: "see me tomorrow, I'll think about it." So I have to go and

see the boss who gets upset, tells the brewer who does it under sufferance...I'm the troublemaker.'[18]

'Brewers think budgets: "There might be this machine that could save me £50k/year, but I'm within my budget so what is the pressure". All he wants to do is brew beer.'[19]

4.8 Status of Energy Management

Case study results suggested that the status of energy management was proportionate to the role of utilities in the cost structure, but interviewees emphasised the repeated conflicts between engineering and brewing staff. Indeed, at four case study sites, this was highlighted as one of the primary barriers to energy efficiency. Engineering and brewing staff had different priorities, different perspectives and different understandings about what was necessary or possible. With brewers having greater authority, this led to a number of conflicts and problems. Examples included competing views about keeping backup boilers on standby and conflicts over repairing a heat recovery system at brewery A.

4.9 Energy Services

None of the case study sites had seriously considered contract energy management and there was scepticism about the value of outside contractors. Risk was an important issue, and interviewees considered that they could not enter into long-term contracts when there was uncertainty over the future of their business. In addition, the overriding concern with reliability led to a preference for having skilled people on site rather than relying on remote monitoring of utilities plant. CHP had been rejected at all sites due to 'peaky' steam demands, excess hot water or other factors which reduced its economic viability.

5. BARRIERS TO ENERGY EFFICIENCY IN UK BREWERIES

As we would expect from a process industry of medium energy intensity, the brewing sector manages energy considerably more effectively than higher education or mechanical engineering. Nevertheless, the survey data on SEC,

Table 5.17 Barriers to energy efficiency in the UK brewing sector

General category	Specific instance	Comments
Hidden costs	Overhead costs of energy management	Salary overheads are significant and may need to be recovered from savings on investment projects. Cuts in staff numbers to reduce salary overheads have resulted in severe time constraints on remaining staff. This provides one of the biggest obstacles to energy efficiency improvements.
	Cost of production disruptions	Can be an obstacle to project and maintenance work. energy efficiency measures must not threaten process reliability.
Access to capital	Availability of capital to the brewery	Borrowing rarely contemplated for small investments. Capital shortage for investment projects within companies, but rationale for restricting borrowing unclear.
	Allocation of capital within company	Other priorities, such as expansion of the retail chain, may squeeze out available funds for energy efficiency investment. Budgets for energy efficiency typically less than best-practice recommendations.
Risk	Business risk	Rationale most commonly used for strict payback criteria. Firms are operating in a difficult business situation, with continuing threat of closures and takeovers.
	Technical risk associated with energy efficient technologies	Any perceived threat to equipment reliability or product quality will prevent energy efficiency investments from going ahead. But this only applies to a subset of energy efficiency opportunities.
Imperfect information	Information on organisational energy use	Many breweries could significantly improve their energy information systems. Level of energy awareness among staff is often poor.

Split incentives	Purchasers of equipment not accountable for energy costs	More relevant to large breweries. Limited adoption of whole-life costing. Problems in integrating energy concerns into purchasing criteria and limited consultation between utilities and project management staff.
Bounded rationality	Decision-making and organisational routines may neglect energy efficiency	Bounded rationality compounds the neglect of energy efficiency encouraged by other factors. In a context of severe time constraints, competing priorities and falling energy prices, energy efficiency is given less attention.

technology adoption and energy efficiency potential, together with the qualitative results from the case studies, suggests that energy efficiency could be cost-effectively improved. By combining the results of the survey and the case studies, we can gain some insights into why breweries impose stringent investment criteria, and why investments that meet these criteria are neglected.

Both the survey and the case study results suggest that not all the technologies listed in Table 5.14 are appropriate for all sites. But once inappropriate technologies have been excluded, there still appears to be a substantial potential for cost-effective energy efficiency improvement that is currently being inhibited by various barriers. These barriers are interpreted below in terms of the theoretical framework developed in Chapter 2. All of the categories of barrier listed in Table 1.1 appear to be relevant and each is discussed in turn below. Table 5.17 identifies the particular ways in which these barriers operate.

5.1 Hidden Costs

Hidden costs may be usefully grouped into three categories: the overhead costs of energy management; costs specific to a technology investment; and loss of utility associated with an energy efficient technology. In the survey, five out of the 21 postulated barriers represented some form of hidden costs. Table 5.18 indicates the respondents' evaluation of the importance of these costs.

The results suggest that the staff time associated with implementing energy efficiency investments was considered very important, while other types of hidden cost were much less important. Each category is discussed below.

Overhead costs

Salary costs form the most important element of overhead costs. If staff numbers are reduced to save salary costs, the time constraints on remaining staff may present an obstacle to energy efficiency improvement. In the survey, time constraints were considered the third most important of the postulated barriers, and were particularly important for small breweries.[20] In the case studies, all interviewees emphasised that time constraints were critical and that they had got substantially worse over the past 5–10 years.

> 'The problems are lack of time, lack of people – which amounts to the same thing – and quality taking precedent. But this is in a 95/5 ratio. Time is the primary issue. I don't consider either capital or information to be a problem.'[21]

> 'In 15 years my staff has reduced from 45 to 25...Margins are tight and resources are stretched. Things are worse for the brewery than for the

company as a whole. So reducing head count is seen as the best way of reducing costs.'[22]

Table 5.18 Importance of hidden costs in the UK brewery energy management survey

General category	**Postulated barrier**	**Overall score**	**% often or sometimes important**	**Ranking in list of 17 barriers**
Overhead	Lack of time/other priorities	1.20	89	3
Utility	Loss of benefit (e.g. unreliable)	0.94	55	10
Specific	Cost of identifying opportunities, analysing cost-effectiveness and tendering	0.81	66	13
Specific	Cost of production disruptions, hassle and inconvenience	0.78	65	14
Specific	Cost of staff replacement, retirement and retraining	0.47	39	17
Mean		***0.84***	***63***	***11***

Note: The average score for the postulated barriers was 0.97.

'There are lots and lots of things where you think "If you looked at so and so that would save us money", but you don't have time to follow it up.'[23]

In both large and small breweries, redundancies had been used to save substantial costs very quickly, and increasing workloads and competing priorities had squeezed out attention to energy efficiency. Core energy management activities had also been affected by the time taken to negotiate supply tariffs in rapidly changing energy markets (rational, in view of the cost savings achieved).

Stringent payback criteria for investment projects could be justified as a means to recover salary overhead costs, but this rationale was never explicit. Many energy management activities were an essential overhead, notably in maintaining reliable operation of utilities plant.

Costs specific to a technology investment

The survey found that costs specific to individual investments were considered less important than general overheads, and few interviewees made specific reference to, for example, the cost of identifying opportunities and tendering. But to the extent that these take up management time, they are subject to the constraints indicated above. Lack of time was a primary reason for the neglect of energy audits.

The cost of production disruptions was mentioned repeatedly by the interviewees, suggesting (contrary to the survey) that this can be important. Disruptions to cask ale production translate directly into an inability to deliver product the next week and a consequent loss of revenue and reputation. To avoid the risk of disruptions, several of the breweries ran backup boilers on warm standby, despite the fact that the lead boiler(s) had demonstrated very high reliability. Similarly, the copper vapour heat recovery at brewery A had not worked for six months due to a stuck valve, but the additional energy costs of £60k/year were considered to be outweighed by the potential loss of production associated with a repair.

Loss of utility

Problems associated with loss of utility scored tenth in the survey, while interviewees emphasised the potential trade-offs between energy efficiency and reliability. Breweries were reluctant to rely on contractors for equipment maintenance, as this may increase the length of production interruptions – a view which biases them against contract energy management.

A second theme was the overriding importance of product quality. Measures were often taken to ensure product quality that had the side effect of increasing energy intensity. A brewery's reputation stands or falls on the quality of its product, so staff were reluctant to take any measures that could threaten this, even if the risk was slight:

> 'Quality is king in brewing...We receive dictates from the Group Quality Assurance Division on things like dissolved oxygen content. Generally there will be no reference to costs and benefits, we simply have to follow. Now that can have major implications for energy use.'[24]

One brewery had a beer recovery process in which the beer was sterilised before being put back in the process at a cost of £15k/year (0.6 per cent of the utilities bill). The facilities manager did not consider that sterilisation was necessary and the preliminary results from a trial had supported his view. But production staff considered that £15k/year was a small price to pay to guarantee product quality, so the sterilisation process was still being used.

Against this it should be noted that many energy efficient investments had additional benefits. For example, changes to defrosting regimes at one brewery led to water savings, reduced maintenance and reduced deterioration of building fabric, in addition to energy savings. In fact, most investments have multiple effects, relatively few are justified on energy savings alone, and quantification of such multiple effects can be difficult.

Loss of utility and potential additional benefits are specific to individual technologies and situations. But it is clear that energy efficiency measures which are perceived as posing a threat to reliability or product quality will not receive a wide take-up in the sector.

Summary

Salary overheads are a very important hidden cost and the resulting time constraints can partially explain the neglect of energy efficiency opportunities. But it is difficult to quantify these overheads, and breweries do not have an explicit requirement to recover overheads from investment projects. Other elements of hidden cost are much less important, but concern over production disruptions and product quality may explain the neglect of some opportunities.

5.2 Access to Capital

From the perspective of survey respondents, limited access to capital provided one of the biggest barriers to energy efficiency, with 'other priorities for capital investment' being rated second in importance. In the case studies, a combination of low brewery profitability, high profitability in the retail and leisure sector and the threat (or opportunity) of acquisitions and takeovers had led to funds being diverted to other areas. There were also competing investment priorities within the brewery itself, such as expansion of production and compliance with legislative requirements on hygiene and safety. The priority given to energy efficiency had fallen during the 1990s following substantial reductions in energy prices.

In addition to capital budgeting, capital was rationed through the use of strict and inflexible investment criteria. Examples were given of where this had prevented projects going ahead, although interviewees considered that time constraints were generally more important.

The rationale for capital rationing and strict paybacks proved difficult to identify. While there was a reluctance to increase borrowing, the main issue seemed to be the strategic priorities of management. For example, one brewery had raised substantial sums for a takeover, but resisted borrowing smaller sums to exploit cost-saving opportunities. This may be because the financial market was expected to react more favourably to the first type of investment than to the second, but the interviews failed to provide a conclusive answer to this question.

In large breweries, most energy efficiency investments involve small sums of money which should have little impact on the overall level of borrowing. But here, bounded rationality and transaction costs become relevant. Energy management staff were preoccupied with working to existing budgets, subject to time constraints and rarely in a position to access external finance. In contrast, financial staff and senior management knew little about the cost-saving opportunities and were preoccupied with business development projects which, unlike cost-saving projects, could not be postponed. Arranging additional finance may have been time consuming and costly compared to the revenue savings from small projects. Put another way, transaction costs impeded both sides from sharing information and reaching decisions that could benefit the firm.

Larger cost-saving projects were typically given more careful scrutiny,[25] but hurdle rates were the same and there was little evidence that such projects were more likely to go ahead. External finance had been seriously considered for CHP, but this had been rejected through a combination of poor economics, unfamiliarity with the technology and suspicion of ESCOs. The result resembled the 'investment gridlock' described by DeCanio (1994, p. 116) where cost-effective projects were not undertaken due to lack of capital, but no outside capital was brought in due to transaction costs and the bias against ESCOs and external funding.

In sum, we can conclude that, from the perspective of survey respondents and case study interviewees, limited access to capital provides a significant barrier to energy efficiency. Explanations for this can be developed with reference to management priorities, the reaction of financial markets to increased borrowing, limitations on individual decision-making and transaction costs. But the rationality and relative importance of each is difficult to judge.

5.3 Risk

Two types of risk are relevant: *business* risk, resulting from broader economic and political trends as well as the specific circumstances of the individual firm; and *technical* risk associated with particular energy efficient technologies.

The survey found that technical risk was considered important (ranking fourth) but business risk much less so (ranking fifteenth), results which were largely independent of brewery size. Interviews found that concerns about technical risk centred on equipment reliability and product quality – issues discussed above as an aspect of hidden costs. This is only relevant for a subset of energy efficiency opportunities. Also mentioned was the risk associated with large investments and with unfamiliar technologies, such as CHP:

> 'We did consider a gas turbine but was a big risk for £2–3 million. Also you'd have to export electricity. Brewers are not generators. We don't understand how the electricity market works. The economics may have been better [than a gas engine]....But £3m compared to £600k is a big difference – a much bigger risk.
> *But surely an ESCO would fund this?*
> Yes, but they would want their profit as well...my bosses don't trust them because what they are into is brewing beer. When we proposed the gas turbine to them, they said "What is a gas turbine?". Well, it is based on an aircraft engine. "A jet engine in a brewery? Sounds like trouble to me. I'd rather buy a few beer tanks or a packaging plant." It was kit they didn't understand.'[26]

The low ranking of business risk in the survey contrasts with the importance given to it by interviewees who emphasised that the current business situation encouraged short time horizons and provided a rationale for strict investment criteria. This applied both to large breweries, where there had been rationalisation and concentration of production at the most efficient sites, and to smaller breweries struggling to compete in an overcrowded market. The importance of business risk may not have been fully appreciated by survey respondents because it is manifest in the strict investment criteria they are required to meet, rather than being faced directly. All interviewees mentioned the risk associated with long contract lengths and a number gave the example of a brewery which had closed while an energy service contract was still in operation.

5.4 Imperfect Information

Three categories of information are relevant – the volume and pattern of energy use in the organisation, the opportunities available for energy efficient investment, and the energy consumption of new equipment. The case studies suggested a correlation between the quality of information systems and the breweries' performance in energy efficiency, but anonymity requirements prevented the survey results from being correlated with SEC data. The survey did demonstrate the scope for improving information systems with, for example, only half of small breweries calculating their SEC.

'Lack of information/poor quality information on energy efficiency opportunities' was ranked ninth in importance in the survey, and interviews suggested that the problem was not so much the availability of information, but the lack of time to read literature and attend workshops. Most interviewees considered that they had sufficient information on energy efficiency opportunities and pointed to time or capital as the more important constraints.

The survey found that a wide range of information sources on energy efficiency were used and there was a wide range of views on their usefulness (Table 5.19). Although several sources were considered good, few were rated as excellent. Interviewees had mixed views on the usefulness of professional bodies and reported that information sharing in the sector had declined as a result of increased competition. A quarter of survey respondents and more than half of the interviewees had never seen the government best-practice literature.

5.5 Split Incentives

Unlike the higher education sector, split incentives did not feature prominently. 'Departments/individuals not accountable for energy costs' was ranked seventh in importance in the survey and while interviewees acknowledged that charging can provide incentives, few considered these to be effective. The consensus view was that the most important changes result from new investment. The scope for savings through improved housekeeping was considered limited, due both to technical limitations and the difficulties of motivating shop-floor staff.

If the staff responsible for purchasing equipment are not accountable for running costs, the energy efficiency of the equipment is likely to be ignored. 'Energy objectives not integrated into operating, maintenance or purchasing procedures' was ranked fifth in importance in the survey, and interviews found that project managers at large sites were accountable for capital budgets but not for whole-life costs. Lack of incentives and lack of consultation with utilities staff led to inefficiencies. The fact that relatively few breweries (29 per cent of survey respondents) had a formal energy policy is important, as this is one means of including attention to energy efficiency within organisational procedures. As an example, brewery C now specifies energy efficient motors as standard in new equipment as a result of its developing environmental policy.

Table 5.19 Sources of information on energy efficiency in the UK brewery energy management survey (per cent of valid replies)

Information source	**Excellent**	**Good**	**Avg.**	**Poor**	**Don't use**	***Weighted score***
Professional organisations	0	56	29	3	12	***141***
Technical conferences/seminars	0	40	40	9	11	***120***
Trade journals	0	29	60	6	6	***118***
DETR Best Practice Programme	6	35	29	6	24	***117***
Equipment suppliers	0	34	40	14	11	***108***
Colleagues within the company	3	37	20	26	14	***103***
Energy manager networks	3	31	20	9	37	***91***
Contacts in other companies	0	27	36	0	36	***90***
Energy supply industry	0	20	43	20	17	***83***
Regional energy efficiency offices	0	14	40	6	40	***68***
Mean	*1*	*32*	*36*	*10*	*21*	***104***

Note: Weights are: excellent = 3; good = 2; average = 1.

In sum, split incentive problems related to accountability and equipment purchasing are evident within the sector – particularly at large sites – but are only considered of medium importance as a barrier to energy efficiency.

5.6 Bounded Rationality

Limitations on individual decision-making are likely to compound the importance of the factors described above. In breweries, we find an overwhelming prevalence of time constraints at all levels combined with an incentive structure dominated by ensuring product quality, meeting production targets and avoiding interruptions. This is set in a context of declining energy prices, a difficult business situation, an uncertain future and tension between brewing and engineering cultures. As in the HE sector, all these factors encourage the development of individual and organisational routines that neglect energy efficiency:

> 'The problem that you've got is the pressure that the industry is under now. We've just had a dramatic cut in labour again...at the end of the day you've got to get the beer out the door...If people are getting pushed and pushed then they're not going to find the time to do the basics.'[27]

> 'I haven't got the buy-in I would like from the small pack area. I still think there are vast amounts of money that could be saved there. But because the whole of their priority is in getting the packaged beer out the door, there's never enough resources to get involved in energy and environment. Such a damn shame.'[28]

> 'Their prime focus is getting the product through and getting the quality right. To be honest, energy efficiency is at the back of their minds.'[29]

Bounded rationality thereby compounds the neglect of energy efficiency that is encouraged by other factors. Its significance may be reduced by biasing decision-making routines towards energy efficient options through either changes in organisational procedures (for example, adopting an energy policy) or public policy (for example, market transformation schemes for electric motors). But while this is feasible for generic technologies such as lighting, it is less appropriate for many of the process-specific opportunities found in the brewing sector.

6. POLICY IMPLICATIONS

There are a range of measures which individual breweries could take to improve their energy efficiency, and there are some excellent best-practice examples which could be followed in both the brewery sector and related sectors (for example, EEO, 1995). These apply particularly to the larger sites, but simplified measures could be adopted by smaller sites. The recommendations have much in common with those developed for the higher education sector, which were summarised earlier in Table 4.18. They include: adopting an energy policy; establishing an effective energy information system; creating an energy management profit centre; including incentive terms in staff contracts; integrating energy efficiency objectives into capital projects and routine maintenance; and using external financing.

For initiatives such as these to be adopted, energy efficiency must become a strategic issue for top management. This is a challenge in the current business situation and the case studies suggested that cost-cutting through redundancies has made it more difficult to improve energy management. But UK and EU climate policy may provide an incentive to change management priorities. This section highlights the role of information programmes and describes how developments since 1999 in UK climate policy are affecting the brewing sector.

6.1 Information Programmes

Brewing differs from the higher education and mechanical engineering sectors in that many energy efficiency opportunities are process-specific. This means that minimum technology standards and market transformation initiatives are less appropriate forms of policy intervention. However, there is scope for expansion and targeting of best-practice literature for the sector, including good-practice guides on brewery processes and case studies of individual projects. Cross-sectoral work on generic processes is also relevant, including compressed air, motors and drives, heat recovery and general energy management. A priority must be the smaller breweries, which have poorer SECs and are more likely to lack information. More effective than literature is a programme of workshops and visits to best-practice sites, which also encourages informal networking. Such programmes face the obstacle of time constraints and reluctance to share ideas with competitors, and hence are only likely to succeed if broader climate policy measures focus management attention on energy efficiency.

Promotion through sector bodies, such as the BLRA and the Brewery Engineers Association, provides a credible and cost-effective method of transferring energy efficiency messages. Government agencies could work with these associations by sponsoring joint workshops, placing articles in relevant journals, advertising programme literature in association publications and including energy efficiency in training programmes. The aim should be to alert both brewing and engineering staff to energy efficiency opportunities to help overcome the cultural divide. Marketing should emphasise the dual benefits of cost saving and environmental improvement.

6.2 The Climate Change Levy (CCL) and Climate Change Agreements (CCAs)

The survey and case studies were conducted in 1999, when UK policy for meeting its obligations under the Kyoto Protocol was at an early stage. Since then there have been a number of important developments, including the introduction of the Climate Change Levy (CCL) and the associated Climate Change Agreements (CCAs). The brewing sector was one of the first to complete a CCA and there is no doubt that the process has led brewing companies to seriously examine their approach to energy management. The cohesiveness of the sector, the comprehensive membership of the BLRA[30] and the attention paid in the past to energy surveys and sector benchmarks have all facilitated the negotiation process. The nature of the agreement and its importance for the sector is briefly discussed below.

Framework

The Climate Change Levy (CCL) is a downstream, revenue-neutral energy tax for business, commerce and the public sector. Gaining political agreement on this tax was a major achievement for the UK government, but it required a series of concessions to be made to energy-intensive industries. 'Facilities'[31] in these sectors have entered into negotiated agreements, termed Climate Change Agreements (CCAs), to give them exemption from 80 per cent of the CCL.

Eligible facilities are those regulated under the EU Integrated Pollution Prevention and Control (IPPC) Directive, which includes 30 of the larger breweries.[32] To comply with competition law, facilities which fall below the IPPC size thresholds are also eligible and have also taken on CCAs – including the majority of small breweries. The CCAs were negotiated between the government[33] and sector representatives and take the form of relative or absolute targets for either energy use or CO_2 emissions, specified at two year intervals up to 2010. The penalty for non-compliance is a return to paying 100 per cent of the CCL for the following two years. In total, CCAs have been negotiated with 44 industrial sectors representing around 6000 industrial facilities (DETR, 2001).

The CCAs vary widely in their choice of base year, the improvement required over a business-as-usual baseline, the assumptions used about production levels and product mix, and the provisions for 'risk management'.[34] In all cases, the targets are based upon a percentage of the 'cost-effective' energy efficiency potential, identified through modelling work by AEAT and described earlier in section 2.5 (ETSU, 2001). Several commentators have argued that the targets are weak, as a consequence of information asymmetry, limited sectoral and technology disaggregation in the GAD database, the restriction to currently available technology, the choice of simple paybacks rather than discounted cash flow for investment appraisal, the very short paybacks used (2 to 4 years) and the fact that only a percentage of cost-effective improvements are required (Sorrell and Smith, 1999; Waller, 2001). In response, industry has emphasised the importance of hidden costs, such as management time and constraints on capital availability (ETSU, 2001).

In addition to the basic agreements, the CCAs incorporate emissions trading arrangements as part of the UK Emissions Trading Scheme (UK ETS) (DEFRA, 2001). These arrangements allow individual CCA facilities to generate 'allowances' if they perform better than their target, and to use allowances for compliance if they perform worse than their target. Sale of allowances is only possible once compliance with the milestone targets has been verified. Allowances can be traded with other CCA facilities and also with the 'direct participants' in the UK ETS. Direct participants are a group of 32 companies from a variety of sectors which have volunteered to take on absolute targets for greenhouse gas emissions in exchange for subsidy payments from the government.

Brewing sector Climate Change Agreement

The target for the brewing sector is a *primary* specific energy consumption of 2.05MJ/ by 2010 (DETR, 2001). This assumes a conversion factor for the fuel used in electricity generation which is fixed from 2000 onwards.[35] The 1999 primary energy SEC was 2.32MJ/l, implying an improvement of 11.6 per cent over the 11 year period. Milestone targets are based on a linear interpolation between 1999 and 2010.

Specification of the targets in terms of primary energy consumption allows the efficiency benefits of self-generation and CHP to be accommodated, but obscures the underlying trend in delivered energy consumption. Assuming no CHP installation[36] and no change in fuel mix, the brewery sector target implies a *delivered* energy SEC of 1.53MJ/l for 2010, which is a 10 per cent improvement on the 1998 delivered SEC (1.70MJ/l).[37] This translates to an average annual improvement over the period 1998–2010 of 0.87 per cent, which compares with an average annual improvement since 1976 of 2.6 per cent, or with an average annual improvement since 1990 of 1.8 per cent. The agreement therefore implies a significant slowdown in the rate of energy efficiency improvement, which may either be because the most cost-effective measures have already been undertaken, or that fewer improvements can be expected from rationalisation and closures, or that the target is weak. The target is based on the projections from the GAD database (discussed earlier) which suggest that adoption of 'all cost-effective' measures would achieve a 14.5 per cent improvement between 2000 and 2010 (ETSU, 2001). The agreement therefore includes around 70 per cent of the estimated 'cost-effective' potential.

In addition to the sector average target, each participating site has an individual target, which is adjusted for production output and packaging mix. This allows companies to avoid penalties if they have met their own target but the sector as a whole has failed to comply.

Issues

The CCL and CCAs represent the most important developments in UK energy efficiency policy since the 1970s and have provided an important stimulus to improve energy management in the brewing sector. Nevertheless, the policy framework raises a number of concerns and is potentially incompatible with the forthcoming EU Emissions Trading Scheme (EU ETS) (Sorrell, 2003). The specific issues include:

- *Efficiency*: Negotiated agreements compare badly with energy taxes (or auctioned emission permits) in terms of economic efficiency (Ekins, 1999). In principle, energy taxes should ensure emission reduction at least cost, internalise the full cost of energy use, provide an incentive for continuous improvement, encourage innovation and raise revenue to provide public goods and services, thereby displacing taxes on labour or income. By using negotiated agreements to exempt sectors from the CCL, the UK has protected the competitiveness of these sectors, but is likely to have raised the overall cost of meeting its Kyoto obligations.
- *Scope*: The criterion for including sectors within a CCA was regulation under IPPC, but this represents only a crude approximation to energy-intensive industry (Hewitt, 1999). Brewing is one of many sectors regulated by IPPC that are not energy-intensive as commonly understood.[38] The requirements of competition law have further expanded the number of facilities covered by the agreements by including those below the IPPC size threshold.
- *Integrity*: The use of relative rather than absolute targets creates the risk that increased output will lead to increased emissions. Industry initially opposed carbon targets as a 'cap on growth', but the inclusion of emissions

trading provisions within the CCAs has weakened this argument. Sectors with energy efficiency targets can participate in the UK ETS, but the provisions to allow this have significantly complicated the design of the scheme (DETR, 2001).

- *Transparency*: The CCAs were negotiated between the DETR and the trade associations with practically no involvement of outside bodies. Even the UK Environment Agency was largely excluded,[39] despite its knowledge of industrial processes and despite its relevant interest as the regulatory body for IPPC. The public and NGOs had limited access to the data and assumptions on which the targets were based, and the complexity of the agreements has made it difficult to interpret the information that has been released (ENDS, 2003).
- *Transaction costs*: The negotiation process proved to be extremely resource intensive, taking up 29,000 person hours of civil servants' and consultants' time (ENDS, 2003). This does not include the effort demanded of trade associations and the 5000-odd companies involved, or the subsequent work in monitoring, verifying and reporting.
- *Stringency*: Negotiated agreements suffer from severe problems of information asymmetry between industry and the regulator (Glachant, 1994). The motivation of the regulator is to determine the abatement costs of the sector and to set a target according to an agreed criteria of costs and benefits, while the motivation of industry is to reduce the stringency of the agreed target by exaggerating abatement costs. Industry has private information on costs which is difficult for the regulator to access without detailed site audits. There is a risk that the agreed targets will be weak and will lead to more costly abatement being required elsewhere. In case of the CCAs, the targets are based on highly aggregated information in the GAD database and assume the use of stringent investment criteria. The brewing sector target implies a significant slowdown in the rate of energy efficiency improvement, which seems unlikely to be justified. In practice, the sector achieved a 6.4 per cent improvement in aggregate energy efficiency at the end of the first milestone period in 2002, which is almost three times larger than the required target for 2002 (2.2 per cent) and approximately equal to that required for 2006 (6.6 per cent) (FES, 2003). In general, the ease with which most CCA sectors have met their first milestone targets suggests that the criticism of weak targets is valid (ENDS, 2003).[40]
- *Trading*: Emissions trading offers CCA facilities a highly cost-effective route to avoiding non-compliance penalties, since the cost of purchasing allowances to cover marginal exceeding of the CCA target is much less than the cost of CCL payments on all fuel and electricity use over a two-year period.[41] This is especially the case in the oversupplied UK market, where surpluses from several of the direct participants have helped to push

UK ETS allowance prices as low as £2/tCO_2 (ENDS, 2002). Errors in establishing emission baselines for the direct participants have led to large volumes of 'hot air' being introduced into the UK market, thereby damaging the credibility of the scheme. By purchasing this 'hot air' and using it for compliance, the CCA companies can avoid taking real action to reduce emissions. Overall the trading arrangements have both increased the incentive for individual facilities to comply with their targets, and provided a cheap and environmentally dubious mechanism with which to do so.

Summary

A complex framework for climate policy has emerged in the UK, based around the CCL, CCAs and emissions trading. These instruments are having a considerable impact throughout UK business and have placed energy efficiency on the agenda of senior management. Breweries are no exception to this, and contacts with the case study sites have suggested that much has changed since the interviews were conducted in 1999. The CCL must take much of the credit for this, and provides a clear example of the effectiveness of environmental taxation in drawing management attention to an issue.

At the same time, the current policy process creates some risks. There are conflicts between the different instruments and between the instrument package and the forthcoming EU ETS, as well as serious concerns about the scope, form, transparency and stringency of the negotiated agreements. In particular, the brewing sector targets imply a much lower rate of energy efficiency improvement than has been achieved over the past 25 years.

7. SUMMARY AND CONCLUSIONS

The UK brewing industry manages energy fairly effectively and has achieved substantial improvements in energy efficiency over the last 25 years – partly as a result of concentration of production at large sites. Nevertheless, performance remains variable and evidence from a range of sources suggests that energy efficiency could be substantially and cost-effectively improved.

The primary obstacle is the overhead costs of energy management and the constraints on staff time. This is a real hidden cost which should be allowed for in appraisals. The brewing industry is ruthlessly competitive, and while margins could be improved through using energy more efficiently, this has taken second place to cost-cutting through redundancies. A second factor is the reluctance to borrow for cost-saving investments and the prevalence of strict and inflexible investment criteria. Explanations for this can be developed with reference to business risk, management priorities, the potential reaction of financial markets to increased borrowing, transaction costs and other factors. But the extent to which this represents rational behaviour as opposed to organisational failure is difficult to judge

Since the bulk of energy use is process-specific, public policies to improve energy efficiency in the sector should aim at encouraging and supporting organisational initiatives. The key to this is to raise the profile of energy efficiency among senior management. The CCL and CCAs have achieved this and have reversed the neglect of energy efficiency by UK industry that has characterised the last decade. It is likely that the negotiated agreements will ultimately be replaced by participation in the EU Emissions Trading Scheme. Whether this increases or dilutes the incentive to improve energy efficiency will depend upon the stringency of the overall (EU wide) cap.

NOTES

1. Case study interview, autumn 1999.
2. The logic of this is demonstrated by the first half results for Bass in 2000. Profits in the hotel and restaurant business were up 21 per cent, those in leisure were up 31 per cent, while brewing profits increased by only 3 per cent. The brewing part of Bass Group was sold to the Belgian company Interbrew in June 2000 for £2.3 billion. Bass is therefore pulling out of the business in which it made its name to become a broad-based leisure company. The situation is not hugely different among the regional brewers, who have experienced a series of takeovers and mergers and who are also concentrating resources on the retail chain.
3. The shift towards lager and small packages has led to electricity growing from 11 per cent of total consumption in 1976 to 21.8 per cent in 1996. In 1989, electricity accounted for 16 per cent of consumption but 45 per cent of costs (DTI, 1994; Central Statistical Office, 1989).
4. This changed its name to the British Beer and Pub Association in 2001.
5. ENUSIM is an engineering model of energy use in the industrial sector. It is based upon a detailed database (the GAD database) of energy consumption by end-use together with energy saving opportunities, disaggregated into 9 main sectors and 60 individual subsectors. The GAD database contains information on around 400 'devices', such as kilns and motors, together with 4000 'technologies' which can be retrofitted to devices, such as energy efficient burners. The database contains capacity data for the base year, performance data and cost data, together with assumptions about 'penetration factors' for individual devices and technologies. The model is driven by exogenous forecasts of fuel prices, investment criteria and demands for physical output from the devices, and projects the most cost-efficient device–technology combinations to meet the forecast output requirements. Increasing output is met through increasing the load factor of existing technologies and installing new capacity when required.
6. In 2002, the Energy Efficiency Best Practice Programme (EEBPP) was renamed 'Action Energy' and taken over by the newly-formed Carbon Trust, while ENUSIM was taken over by Cambridge Econometrics. Ownership of ENUSIM resides with the government.
7. Brewing accounts for 70 per cent of energy consumption in this combined sector.
8. This is *primary* specific energy efficiency, which includes conversion factors for electricity generation. Since the conversion factors are fixed for the projection period, and since no increase in CHP capacity is projected (due to lack of information), the percentage changes are the same for delivered energy efficiency.

9. It was not possible to link the results of the energy survey with the BLRA data as this would violate the anonymity requirements of the survey.
10. A non-valid reply means that no answer was given.
11. See http://www.macc2.org.uk.
12. Monitoring and Targeting (M&T) is the UK term for a comprehensive energy data collection and analysis system. M&T is supported by a variety of software packages (Croner, 2000).
13. Energy-specific investments were considered more important in large breweries (58 per cent), while investments undertaken for other reasons were considered more important in small breweries (64 per cent).
14. Best-practice literature recommends that organisations employ one full-time utilities manager for each £1 million of annual utilities expenditure (that is, energy + water). Assuming salary costs of £30k/year, this equates to salary overhead costs of ~3 per cent of the utilities bill. This guideline can be adjusted pro-rata to identify an appropriate commitment of staff time for an organisation. However, there are considerable practical difficulties in identifying which activities count as energy management and in estimating staff time when activities are divided between several people.
15. Performance is assessed here on the basis of SEC figures, the percentage of technologies adopted and qualitative judgements.
16. For example, one company had adopted a target of purchasing a certain number of pubs a year.
17. UK brewery case study interview, autumn 1999.
18. UK brewery case study interview, autumn 1999.
19. UK brewery case study interview, autumn 1999.
20. The difference in weighted average scores between small and large breweries for this barrier was three times the survey average.
21. UK brewery case study interview, autumn 1999.
22. UK brewery case study interview, autumn 1999.
23. UK brewery case study interview, autumn 1999.
24. UK brewery case study interview, autumn 1999.
25. At brewery E, approval was relatively straightforward for projects of < £10k, while approval from a committee was required for projects in the range £10–100k. Projects >£100k required detailed investigation from a Group project management team.
26. UK brewery case study interview, autumn 1999.
27. UK brewery case study interview, autumn 1999.
28. UK brewery case study interview, autumn 1999.
29. UK brewery case study interview, autumn 1999.
30. Fifty out of 56 companies.
31. A facility comprises one or more IPPC installations and may also include other activities. For example, where an installation (or group of installations) consume more than 90 per cent of a site's energy use, then all of the energy use at the site will be covered by a CCA.
32. As sites in the food and drink sector producing >300 tonnes/day.
33. Relying heavily on the expertise and sector-specific knowledge of AEA Technology.
34. Some sectors are allowed to adjust their targets if there are changes in product mix or output level, while others have adopted a 'tolerance band' around their target.
35. The conversion factor for delivered electricity to primary fuel use is 3.2 for 1990, 2.68 for 1999 and 2.60 for 2000 onwards.
36. The agreement assumes no increase in CHP capacity owing to insufficient site by site knowledge. It requires an assessment of CHP potential within the first two years, followed by an adjustment of the targets if appropriate.
37. BLRA, personal communication.
38. A commonly accepted definition of 'energy-intensive' is where energy costs account for more than 10 per cent of total business costs.
39. UK Environment Agency, personal communication.
40. Overall, CCA facilities reduced emissions by 15.8MtCO_2/year below the baseline, or 13.5MtCO_2/year below an 'equivalent' 2000 baseline (FES, 2003). This is more than three times the cumulative target for the first milestone period (2002) and significantly greater

than the final target for 2010 (9.2$MtCO_2$). Some 70 per cent (9.5 $MtCO_2$/year) of this was contributed by plant closures and output reductions in the steel industry, but the rest of industry reduced emissions by 4$MtCO_2$/year, or 25 per cent more (1$MtCO_2$) than required by the first milestone period.

41. As an illustration, assume a firm with an annual electricity consumption of 1GWh. Its CCL liability for electricity is €6020. Assume it agrees to a CCA with a target of 10 per cent saving in electricity use (100MWh) in return for a CCL exemption worth €4816 (80 per cent of €6020). Assume it only achieves a 5 per cent improvement (50MWh). To make up the shortfall, the company needs to buy allowances for 50MWh or 21.5tCO_2 (assuming a carbon intensity of delivered electricity 0.43tCO_2/MWh). At an allowance price of €5/tCO_2, that would cost €108. In contrast, the cost of losing the exemption for two years would be 111 times greater at €12040.

REFERENCES

BECOM (1995), *Brewing and Malting: Economy through Energy Efficiency*, Thermie Programme Action No. 1251, DGXVII, European Commission, Brussels.

Brewers and Licensed Retailers Association (1997), *Statistical Handbook,* Brewing Publications Ltd., London.

Brewers and Licensed Retailers Association (1998), 'Technical circular no. 319: brewery energy survey report 1976–1996', London.

Brewers and Licensed Retailers Association (1999), 'Technical circular no. 344: utilities and environmental survey report 1998', London.

Central Statistical Office (1989), *Business Monitor PO1008: Purchases Inquiry*, HMSO, London.

Croner (2000), *Croner's Energy Management Guide*, Croner CCH Group Ltd, Kingston, Surrey.

DeCanio, S.J. (1994), 'Agency and control problems in US corporations: the case of energy efficient investment projects', *Journal of the Economics of Business*, **1**(1), 105:123 .

Department of the Environment, Food and Rural Affairs (2001), *The UK Emissions Trading Scheme*, HMSO, London.

Department of the Environment, Transport and the Regions (2001), *Umbrella Climate Change Agreement for the Brewing Sector*, London.

Department of Trade and Industry (1994), *Industrial Energy Markets: Energy Markets in UK Manufacturing Industry 1973–1993*, HMSO, London.

Ekins, P. (1999), 'Environmental taxation, equity and competitiveness', a paper for the Government's roundtable on sustainable development, Warwick University, June 25.

Energy Efficiency Office (1992), *Energy Consumption Guide No. 29: Small Breweries*, ETSU, Harwell.

Energy Efficiency Office (1993), 'Organisational aspects of energy management', General Information Report 12, ETSU, Harwell.

Energy Efficiency Office (1995), 'Monitoring & targeting at a brewery', Good Practice Case Study No. 273, ETSU, Harwell.

Environmental Data Services (2002), 'Hot air blows gaping hole in emissions trading scheme', *The ENDS Report,* March, **326.**

Environmental Data Services (2003), 'Peering through the murk on the Climate Change Agreements', *The ENDS Report*, April, **339**.

ETSU (1996), 'Industrial sector carbon dioxide emissions: database and model for the UK', report to the DETR Global Atmosphere Division, Ref RYTA 18616001/Z/6, ETSU, Harwell.

ETSU (1999), 'Industrial sector carbon dioxide emissions: projections and indicators for the UK 1990–2020', report to the DETR Global Atmosphere Division, ETSU, Harwell.

ETSU (2001), 'Climate Change Agreements – sector energy efficiency targets', Version 1, ETSU, Harwell.

European Commission (1994) *Brewing and Malting: Economy through Energy Efficiency*, a Thermie programme action, DGXVII European Commission, Brussels.

Future Energy Solutions (2003), *Climate Change Agreements: Results of the First Target Period Assessment*, AEA Technology, Harwell.

Glachant, M. (1994), 'The setting of voluntary agreements between industry and government: bargaining and efficiency', *Business Strategy and the Environment*, **3**(2), 43–9.

Harris, P.S. (1979), *Energy Audit Series No. 8: the Brewing Industry*, issued jointly by the Department of Energy and the Department of Industry, HMSO, London.

Hewitt, C. (1999), 'Evidence to the Trade and Industry Select Committee inquiry into the Climate Change Levy', Institute for Public Policy Research, London, 11 June.

Howarth, R.B., B.M. Haddad and B. Paton (2000), 'The economics of energy efficiency: insights from voluntary participation programmes', *Energy Policy*, **28**(6-7), 477–86.

Sorrell, S. (2003), 'Back to the drawing board: implications of the EU Emissions Trading Scheme for UK Climate Policy', SPRU – Science and Technology Policy Research, University Of Sussex, Brighton, available from: http://www.sussex.ac.uk/spru/environment/research/drawingreport.pdf.

Sorrell, S. and A. Smith (1999), 'Avoiding a policy glut over industrial energy use: IPPC, the Climate Change Levy, negotiated agreements and carbon trading', a report to the Environment Agency, SPRU – Science and Technology Policy Research, University of Sussex, Brighton.

Trade and Industry Committee (1999), 'Impact on industry of the Climate Change Levy', Ninth Report Session 1998–99, HC 678, House of Commons, HMSO, London.

United Nations Environment Program: Industry and Environment (1996), 'Technical Report No.3 44, *Environmental Management in the Brewing Industry*, UNEP, Paris.

Waller, D. (2001), 'The Climate Change Levy and negotiated agreements', Discussion paper DP05, Association for the Conservation of Energy, London.

6. Production Must Go On: Barriers to Energy Efficiency in the Irish Mechanical Engineering Industry[1]

Eoin O'Malley and Sue Scott

'…there are no conflicts of interest, except that production must go on.'[2]

1. INTRODUCTION

The mechanical engineering sector encompasses a wide range of industries, extending from miniaturised precision components to complete manufacturing plants. Over half of the products manufactured by the sector are finished capital goods (for example, machine tools, furnaces, engines and so on), while the remainder consist predominantly of components for capital goods. Competitiveness rests on innovation and product quality, with product price of secondary importance. Energy forms less than 2 per cent of total costs on average and is primarily used in generic building applications such as heating and lighting. In addition, the sector is dominated by small and medium-sized enterprises (SMEs) which lack both staff resources and technical expertise in energy management. Taken together, these factors create the perfect context for neglecting energy efficiency opportunities. But while such neglect may be understandable from the perspective of individual firms, it has important consequences for national climate policy. For example, in the UK the 'mechanical engineering and metal products' sector accounts for around 5 per cent of total industrial energy use, which is ten times more than the UK brewing sector.

This chapter explores the extent to which Irish mechanical engineering firms are neglecting opportunities to improve energy efficiency and the validity of competing explanations for this. The study is based upon seven case studies of energy management practices within individual firms. The chapter begins with an overview of the Irish mechanical engineering sector, including the structure of the sector, recent economic trends and the pattern of energy use. The case study firms are then introduced, simple data on their energy efficiency performance is presented, and the relative importance of different barriers to energy efficiency is assessed. This leads to recommendations on how these barriers may be overcome through action at the firm, sector and national level. It is argued that the diversity of barriers and the significance of management time as obstacles point to the need for a mix of policies that can 'short-circuit' individual organisational decision-making.

2. CHARACTERISING THE MECHANICAL ENGINEERING SECTOR

2.1 Features of the Mechanical Engineering Industry

The mechanical engineering industry makes capital equipment for use in manufacturing, mining, energy, agriculture and many types of services.[3] The manufacture of machinery is often referred to as 'old' technology as many of its products were invented over a century ago. But this overlooks the fact that each product is developed continuously, with a typical innovation cycle of 4–5 years. Modern machinery successfully combines 'old' and 'new' technologies, with the incorporation of microelectronics, optics, sensors and new materials. Machinery customers often require tailor-made solutions to their production tasks, which means that the series production of standardised products is rare. Instead, the sector is characterised by a high degree of specialisation by individual producers which offers good opportunities for small and medium-sized enterprises. The greater part of machinery and equipment output is produced individually or in small batches, because this is the only way the customers' varied requirements can be met.

Most machinery and equipment requires considerable design effort and precision in production. A high proportion of the sectors' employees are therefore highly skilled engineers and craftspersons. Input supplies to the sector constitute nearly 60 per cent of the value of its output, and these are dominated by the products of other companies within the sector. Since the industry is dependent upon investment in capital goods by other parts of manufacturing, it tends to experience sharp cyclical fluctuations.

The most important basic material for the manufacture of machinery and equipment is steel, although a large proportion of this is procured in processed form, such as forgings. The prevalence of steel throughout the sector means that there is less diversity in industrial processes than in final products. The primary process used throughout the industry is assembly, but there is also metal cutting and some more energy-intensive processes such as forging, welding, heat treatment, painting and stoving.

2.2 The Irish Mechanical Engineering Industry

The mechanical engineering sector in Ireland employed 11,122 people in 1998 and had a gross output valued at £997 million.[4] The sector accounted for 4.6 per cent of total Irish manufacturing employment and 2.1 per cent of total manufacturing gross output (CSO, 2000).[5] This meant that mechanical engineering made up a smaller share of Irish manufacturing than in the EU as a whole.

There were 344 enterprises in the Irish industry in 1998, and 353 individual plants or factories. This implies that very few of the enterprises operated more than one factory. However, foreign-owned multinational companies accounted for 17 per cent of the industry's factories, 46 per cent of its employment and 57 per cent of its gross output. This degree of foreign ownership is not unusual in Ireland, since foreign firms account for a similar proportion of total manufacturing (Barry et al., 1999). The foreign-owned firms had an average employment of 103 compared to 25 in Irish indigenous firms. Overall, 34 per cent of employment was in small establishments with less than 50 people, 30 per cent in medium-sized establishments with 50–199 employees, and 36 per cent in large establishments with 200 or more employees.

The output of the Irish mechanical engineering industry is spread across a wide range of product categories, with no major specialisation in any particular category. However, compared with larger economies such as Germany or the UK, there are some substantial gaps in the product range (for example, tractors, combine harvesters and bulldozers). Among the more important components of the Irish industry are tool-making, pumps and compressors, lifting and handling equipment, non-domestic cooling and ventilation equipment, and agricultural machinery. The manufacture of machine components, together with activities such as installation, repair and maintenance, constitute a significant proportion of the industry's sales.

The Irish mechanical engineering industry is highly export-oriented, with 72 per cent of its production being exported in 1998 (CSO, 2000). The fact that the industry exports such a high proportion of its output suggests that many firms must be internationally competitive.

The output of the industry increased by 5.8 per cent per year in real terms over the period 1991–98, which was below the average growth rate for all Irish manufacturing, but above the average growth rate for industrial production in the EU. From the early 1960s onwards, the majority of the industry's growth was a result of new investment in Ireland by foreign-owned multinationals. This changed in the 1990s, and the Irish-owned branch of the industry has grown faster in recent years. In 1991–98, employment in foreign-owned firms decreased by an average of 0.8 per cent per year, whereas employment increased by 5.4 per cent per year in Irish-owned firms.

2.3 Energy Use in the Mechanical Engineering Industry

In 1998, the mechanical engineering industry spent £10.8 million on fuel and electricity, which amounted to 1.1 per cent of the value of its gross output and 2.1 per cent of its purchases of industrial inputs. For comparison, expenditure on fuel and electricity amounted to 2.2 per cent of purchases of industrial inputs by all Irish manufacturing, indicating that mechanical engineering was slightly less energy-intensive than the average. Electricity accounted for 56 per cent of total energy expenditure in the mid-1990s, followed by petroleum oils (30 per cent) and natural gas (5 per cent). The share of natural gas has been increasing.

Table 6.1 Trends in energy purchases in the Irish mechanical engineering sector

Year	Energy expenditure (£ million)	% of value of gross output	% of industrial inputs
1985	6.6	1.8	3.5
1990	7.0	1.4	2.6
1994	7.2	1.1	2.2
1998	10.8	1.1	2.1

Note: 1985, 1990 and 1994, data refer to NACE 70 sector 32, 'mechanical engineering', but data for 1998 refer to NACE Rev 1 sector 29 minus 297, 'machinery and equipment n.e.c.' minus domestic appliances.

Source: Central Statistics Office (2000).

Information on the breakdown of energy consumption by end-use for this sector is relatively poor. As an illustration, Table 6.2 presents some UK data for the engineering sector as a whole (that is, mechanical, electrical, electronic and vehicles). This shows that non-production energy use (space heating, lighting and so on) accounts for 55 per cent of total consumption and 60 per cent of fossil fuel consumption. Box 6.1 elaborates on the nature of energy use within each category.

Table 6.2 Estimated breakdown of energy consumption within the UK engineering sector (per cent of total)

Category	**Electricity**	**Fossil fuel**	**Total**
Production			
Forming	25	4	12
Finishing	17	36	29
Assembly/test	1	0	1
Ancillary services	11	0	4
Sub-total	***54***	***40***	***45***
Non-production			
Space heating	4	58	38
Lighting	17	0	6
Air conditioning	8	0	3
IT	7	0	3
Miscellaneous	4	2	3
Ventilation	6	0	2
Sub-total	***46***	***60***	***55***
Total	**100**	**100**	**100**

Source: ETSU, private communication.

Establishing benchmarks for energy efficiency is difficult for such a heterogeneous sector. GJ/tonne of output has little meaning, owing to the huge diversity of products, while GJ/employee is little better since many end-uses will be insensitive to employment levels (for example, labour may be cut with no reduction in the energy required to heat the building). GJ per unit of value added is also problematic, since energy forms only a fraction of total costs and value-added may fluctuate for other reasons. The UK Energy Efficiency Office has calculated some relatively crude benchmarks based on floor area, as shown in Table 6.3.

Box 6.1 End-use of energy in the mechanical engineering sector

Production

- *Forming:* This includes metal forming, machining and joining. Most of the energy is in the form of electricity to drive electric motors and produce compressed air.
- *Finishing:* This includes heat treatment, surface preparation, coating, painting and drying. These use a variety of processes including high

temperature furnaces, medium temperature vats, drying ovens and infra-red heating. Each uses a mix of fossil fuels and electricity and some can be energy-intensive.

- *Assembly:* This uses very little energy beyond electricity for conveyors and transportation.
- *Ancillary services:* This includes compressed air and pumping, which may account for up to one-fifth of electricity consumption.

Non-production

- *Space heating*: This is normally the largest single energy consumer. Many engineering buildings have high ceilings and require the heating of large volumes of air.
- *Lighting*: Most factories now have some form of efficient lighting. But conventional fluorescents are still common.
- *IT and air conditioning*: These are increasing in importance.
- *Ventilation*: There has been an increase in the number and capacity of ventilation systems to meet health, safety and environmental standards, particularly on solvents.
- *Miscellaneous*: This includes canteens and domestic hot water.

Table 6.3 Energy benchmarks for the engineering sector and general manufacturing (kWh/m^2/year)

	Average number of shifts	**Buildings energy**	**Process energy**	**Buildings and process energy**
Engineering	1.7	302	85	387
General manufacturing	3.0	409	495	904

Source: Energy Efficiency Office (1993).

As regards opportunities to improve energy efficiency, there is little formal documentation from Irish sources that is specific to mechanical engineering. However, based largely on UK sources, we identified a list of 23 recommended measures for improving energy efficiency that would commonly be appropriate for mechanical engineering firms. These are listed in Table 6.4.

Following further consultation with the Irish Energy Centre (now Sustainable Energy Ireland), we also identified a shorter list of seven energy efficiency measures that were considered particularly important. These were:

Table 6.4 Selection of energy efficient technologies/techniques for the mechanical engineering sector

Area	Technology/technique
Space heating	Use of Building Energy Management System (BEMS)
	Programming HVAC controls to match occupancy patterns and/or outside temperature
	Thermostatic radiator valves
	Ensure adequate lagging/insulation
Lighting	Use of 26mm compact fluorescents
	Use of high frequency electronic ballasts
	Photocell, acoustic or movement sensors
Compressed air	Inspection & elimination of leaks
	Generation of compressed air at appropriate pressure
	Consideration of energy consumption when purchasing new compressors
	Heat recovery
Heat treatment	Furnace insulation
	Furnace heat recovery
	Accurate control of temperature, pressure and fuel air ratio
Boiler plant	Insulation of distribution pipes, valves and flanges
	Use of energy efficient burners
	Use of condensing boilers
Building fabric	Retrofitting insulation to walls and roofs
	Reduction of draughts from loading bays
	Specification of high standards of energy efficiency in new buildings
Electrical equipment	Purchase of energy efficient computers and office equipment
	Use of variable speed drives (VSDs) for motors
	Automatic switch off of fans & pumps when the equipment they serve is not in use

- installation of a Building Energy Management System (BEMS);
- conduct of a detailed energy audit;
- purchase of energy efficient equipment;
- programming of heating and ventilation controls to match occupancy patterns and outside temperature;

- replacement of tungsten filament lamps with slim or compact fluorescents or high pressure sodium lighting;
- specification of high standards of energy efficiency in new and refurbished buildings; and
- use of variable speed drives in pumps, fans and other applications.

This shortlist was taken into consideration in our study, as a further yardstick when assessing the extent to which firms had adopted energy efficiency opportunities.

3. CASE STUDIES OF ENERGY MANAGEMENT IN IRISH MECHANICAL ENGINEERING

3.1 Selection of Cases

Case studies were undertaken of energy management practices within seven Irish mechanical engineering firms. These were all branches of multinational companies and each employed more than 50 people. The products manufactured included dies, tools, pumps, lifting and handling equipment, engine components and machinery for agriculture and specific manufacturing industries.

The initial approach was made by phone to identify the person having most responsibility for energy matters and to ask if the company would participate. There was usually no single person with the job title of energy manager and individuals with the relevant responsibilities included the technical manager, maintenance manager, safety and environment manager, production manager and chief executive. Energy was very much a secondary issue for all the firms and the division of responsibility between different individuals was frequently unclear. If the company agreed to participate, the relevant individual was sent a 'pre-interview questionnaire' (PIQ), containing general questions on energy consumption, technology adoption and perceptions of barriers to energy efficiency. This was followed by semi-structured interviews of key individuals, administered by telephone (Annex B). Between one and three people were interviewed in each company (Table 6.5).

Table 6.5 Number of persons interviewed in each case study, their job titles and backgrounds

Case study code number	Number of interviewees	Job titles (and background)
1	2	Maintenance Manager (electrical engineer); Managing Director
2	2	Maintenance Supervisor (engineer); Accounts Manager
3	3	Health Safety and Environment Manager (engineer, plus safety and environmental management qualification); Technical Manager; Operations Manager
4	1	Electrical Maintenance Manager (electrician)
5	1	Engineering Manager (engineer),
6	2	Works Manager; Manufacturing Engineer (engineer)
7	3	Quality Manager (engineer); Purchasing Officer; Operations Manager

3.2 Description of the Case Study Firms

Before describing the seven case studies, it is interesting to note that there was a consistent pattern in the firms' responses relating to implementation of energy efficiency measures and their perceptions of barriers to energy efficiency. Firms were ranked according to the number of energy efficiency measures they had implemented, as recorded in the PIQ, and also according to the number of measures implemented from the shortlist of seven most important measures. Of course, this is a little arbitrary since certain measures might not be applicable to all the firms. The firm that implemented most measures was ranked Firm 1, the firm implementing the second most was ranked Firm 2 and so on, as shown in the first and subsequent two columns of Table 6.6. These rank numbers are used as the codes for the case studies from here on.

The ranking can be compared with firms' self-assessment of their management of energy in the fourth column, and with the number of barriers to energy efficiency that they perceived as 'often important', in the fifth column (Table 6.6). The ranking can be further compared with their perception of whether or not there were energy efficiency opportunities available with paybacks of less than three years, in the final column.

Table 6.6 Ranking of firms by number of measures implemented, and firms' perceptions of their energy management, barriers to energy efficiency and opportunities

	Number of measures implemented		Self-assessed profile	Number of barriers 'often important'	Opportunities exist with payback <3yrs
	From shortlist	From PIQ list			
Firm 1	6.5	10	2+	2	Disagree
Firm 2	5	12.5	2	5	Disagree
Firm 3	4.5	12	1+	3	Agree
Firm 4	4	7	1-	12	Neutral
Firm 5	2	4	0+	5	Agree
Firm 6	2	1.5	0+	12	Agree
Firm 7	0(1)	0(2)	1	12(8)	Agree

Notes:

a In columns 2 and 3, a score of 0.5 is given if a measure has only been partially implemented.

b The self-assessed profile refers to the average score from the energy management matrix reproduced in section 6 of Annex A. This ranges from 0 (poor) to 4 (excellent).

c In the case of firm 7, two different entries are given in some cells, reflecting the fact that the PIQ was filled in by two different people.

A few interesting points emerge from the table. Twenty-three possible barriers to energy efficiency were listed in the PIQ (see Table 5.15) and respondents were asked to grade them as 'often', 'sometimes' or 'rarely' important. It can be seen from the fifth column that in general more barriers were perceived to be important by firms that had implemented the least number of energy efficiency measures. In addition, the last column shows whether or not firms agreed that they were missing potential opportunities for energy efficiency improvements that would pay back in less than three years. The adopters of most energy efficiency measures (at the top of the ranking) indicated that there were few remaining opportunities, while the non-adopters stated that there were untapped opportunities. This pattern of responses is reasonable and consistent.

The fourth column represents each firm's self-assessment of its energy management practices, scored from 0 (poor) to 4 (excellent). Here again there is a consistent pattern, with the firms that had adopted the most energy efficiency measures having the highest score. This analysis is of course subjective and imprecise. But the general pattern of responses does at least show an encouraging degree of rationality and consistency on the part of the respondents. Table 6.7 shows that the case study firms employed a total of

1431 individuals, which corresponds to 13 per cent of the sector's total employment. Sales by the case study firms amounted to £142 million, representing 14 per cent of the sector's gross output. Total expenditure on energy by the firms, at £1.5 million, amounted to 1.07 per cent of their value of sales, which compares with the figure of 1.1 per cent of gross output for the sector as a whole. Electricity accounted for 78 per cent of total energy costs on average, with the proportion being lower in the better performing firms.

A notable point from Table 6.7 is the small size of the energy bill compared to the other organisations studied within this project. Universities, for example, typically have an annual bill exceeding £1 million. Of the seven mechanical engineering firms studied, firm 3 had the largest annual bill at £500k, while firm 4 had the smallest at £73k. In the case of firm 4, the annual energy bill was not much greater than the salary costs of a single high-ranking employee. When utility bills are this small, staff costs for energy management activities become proportionately more significant.

Table 6.7 Firms' employment, sales and energy expenditure

	Number of employees	Sales £m/year	Energy expenditure £k/year	Electricity (%)	Fuel (%)
Firm 1	285	20	190	79	21
Firm 2	326	21	297	66	34
Firm 3	180	20	500	50	50
Firm 4	125	25	73	86	14
Firm 5	138	19	96	89	11
Firm 6	165	6	81	86	14
Firm 7	212	31	280	89	11
Total	1431	142	1516		
Average	204	20	217	78	12

It can be seen from Table 6.7 that firms 1–5 have similar sales, in the range £19 million to £25 million, and that firms 4, 5 and 6 have the lowest energy bills. In terms of employment the largest firms are the first two, and their sales are similar. Firm 1, however, uses considerably less energy, partly due to the fact that it contracts out its heat treatment. This in itself could be an energy efficiency measure, if the outside contractor is specialised and efficient. Firm 2 had the most employees and the second highest energy bill, but it considered that 'energy is not an issue, wages are what matters', adding that the overriding consideration was that 'production must go on'. However, its record on adoption of energy measures puts it high in the ranking.

Firm 3 had taken on a safety and environment manager the previous year and its high energy bill had stimulated efforts to lobby for cheaper fuel and undertake energy efficiency improvements. Firm 4 had the lowest number of employees and the lowest energy expenditure. Accordingly only low cost energy efficiency measures were undertaken and a three year payback criterion was applied on the grounds that if it were longer the 'machines could be out of date'. In Firm 5 the driving need for production to go on was underlined by the award of bonuses related to production and machine uptime, but not for energy efficiency improvements. Firm 6 was subject to foreign customer pressure to ensure that its processes met high standards of health and safety, but comparable pressure to improve energy or environmental performance appeared to be lacking.

Finally, firm 7 was an outlier. It had the highest level of sales with possibly the lowest adoption of energy efficiency measures. However it was in the initial phase of ISO14001 certification, on the instruction of its head office which had become conscious of public image and the possible consequences for share prices. Firm 7 was in a good position to explain the difficulties of this type of administrative reform, which entailed establishing an entirely new structure of energy and environmental management.

Table 6.8 adds further information about energy management practices. The pattern is again suggestive of consistent behaviour, particularly in relation to monitoring energy use. The firms that have implemented most energy efficiency measures tend to be those that monitor trends in energy use. The results also suggest that metering at the building (rather than site) level, monitoring and targeting and use of benchmarks are not widespread practices in the sector, but those firms which undertake these tasks tend to perform better.

A point worth noting is that respondents had some difficulty in reporting the quantities of energy used. Figures on energy expenditure were fairly easily obtained, but only one firm reported all the quantities. Another firm's respondent had to contact the finance division to obtain fuel quantity and the remaining firms gave patchy information at best. While there may have been some reluctance to enable outsiders to calculate the unit energy price, the impression given by the respondents is that they were only aware of the energy bill. Because of their uncertain nature, quantity figures and energy intensities are not recorded here.[6]

Table 6.8 Energy information systems in the case study firms

	Monitor trends	Electricity metered at	Use monitoring and targeting	Use benchmarks
Firm 1	Yes	Site	No	No
Firm 2	Yes	Bldg+equip	Yes	No
Firm 3	Yes/No	Bldg	No	Yes
Firm 4	Yes	Site	No ans	No
Firm 5	No	Bldg	No	No
Firm 6	No	Site	No	No
Firm 7	Yes	Site	No	No

3.3 Summary of Case Studies

As a group, the participating firms are a satisfactory sample. They show a good degree of consistency in their responses to the pre-interview questionnaire and their aggregate profile indicates that they are reasonably representative of the mechanical engineering sector as a whole (albeit excluding firms with less than 50 employees). The firms revealed a wide range of performance on energy efficiency, with some having undertaken most of the measures on the shortlist, while others had barely implemented any. Similarly their perceptions of the number of barriers that were important ranged from very few to about a dozen. Over half of the firms felt that there were energy efficiency opportunities available with paybacks of less than three years, while the three firms ranked at the top of the list said that there were few remaining opportunities.

A fairly coherent picture has emerged of implementation of energy efficiency measures, of perceptions and of behaviour. The next task is to see what this tells us about each of the hypothesised barriers to energy efficiency. This is the subject of the next section.

4. BARRIERS TO ENERGY EFFICIENCY IN IRISH MECHANICAL ENGINEERING

We have seen that most firms felt that there were untapped opportunities for improving energy efficiency. This section explores the reasons for this neglect with the help of some simple graphs. These list firms on the horizontal axis, ranked in the order of adoption of energy efficiency measures as before, while their views on the importance of each barrier are represented on the vertical axis using a scale of: 1 = rarely important; 2 = sometimes important; and 3 =

often important. It is argued that the most important barriers are access to capital, hidden costs and imperfect information.

4.1 Access to Capital

Access to capital was considered the most important barrier to cost-effective investment in energy efficiency, and yet access to capital did not appear to be a problem for the firms in general. Firms generally had access to funds, whether internal or borrowed, but an investment could still be barred due to other priorities taking precedence. In many cases, strict adherence to capital budgets was a problem, especially in view of the low priority given to energy efficiency.

All of the firms bar firm 5 were able to borrow for capital investment in general although they tended to favour using internal funds. If an investment was large they might require approval from head office which, with some exceptions, seemed to exercise fairly relaxed control where decisions about energy efficiency were concerned. Despite this, lack of capital and other priorities for investment were still considered to be important, as Figure 6.1 illustrates.

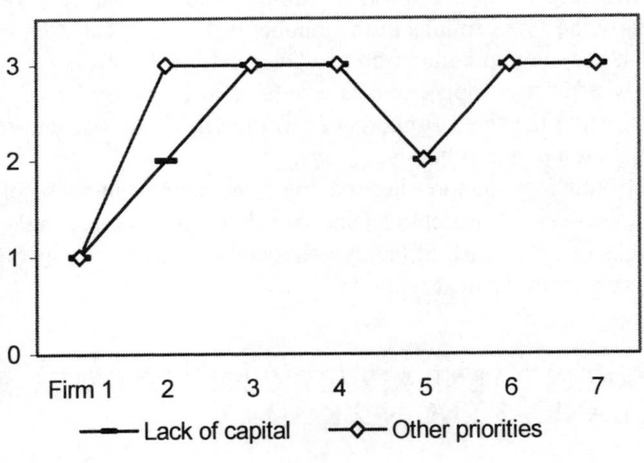

Note: on the vertical axis, 1 = rarely important, 2 = sometimes important, 3 = often important.

Figure 6.1 Importance of lack of capital and other priorities for investment, as barriers to energy efficiency improvement

appraising investment opportunities of this type, with the length of payback required varying between two and five years. One firm used 2.5 years but calculated the NPV for large investments, while firm 1 simply stated that if an investment produced a 'good' return it would be considered. In firm 3, some investments were required to show a 15 per cent rate of return. At least two firms operated both a payback period and a fixed investment budget.

Application of such short payback criteria appeared to rule out a number of projects with good NPVs. Furthermore, in cases where the investment budget was fixed, projects which met the payback criteria were nevertheless squeezed out by other projects that were deemed 'essential'. The latter category included a number of 'environmental' projects, which were either imposed by regulatory requirements or considered necessary to avoid a negative public image for the firm.

In theory, at least, third-party finance could be used to overcome barriers arising from strict payback rules or relegation by other priorities vying for fixed capital budgets. But none of the firms had considered third-party finance and their relatively small size appeared to have put them below the threshold for viable energy service contracts.

4.2 Hidden Costs

Firms were asked to evaluate the importance of six categories of hidden cost:

- cost of production disruptions, hassle and inconvenience;
- cost of identifying opportunities, analysing cost effectiveness and tendering;
- cost of staff replacement, retirement and retraining;
- possible poor performance of equipment;
- difficulty and cost of obtaining information on the energy consumption of purchased equipment; and
- lack of time and other priorities.

Looking at two of these categories at a time, the following three graphs illustrate the firms' views about their importance. Figure 6.2 shows that over half of the firms perceived 'identifying opportunities, analysing cost effectiveness and tendering' to be often important, while 'disruptions, hassle and inconvenience' was at least sometimes important to all but one firm. Hence, the indications are that these two barriers matter.

It is again worth noting the general pattern of responses. The firms that have adopted most energy efficiency measures attach less importance to these barriers. This could be an example of how the fulfilment of a task can leave

barriers. This could be an example of how the fulfilment of a task can leave one feeling that it was not so difficult, or alternatively these firms may have organisational routines which facilitate the adoption of such technologies.

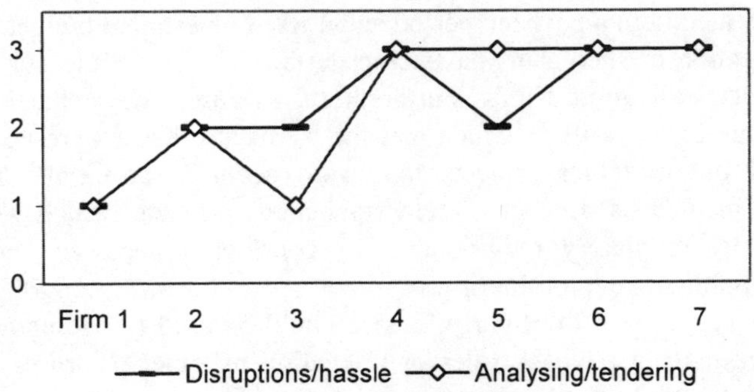

Note: On the vertical axis, 1 = rarely important, 2 = sometimes important, 3 = often important.

Figure 6.2 Hidden costs – importance of disruptions/hassle and costs of analysing and tendering

Figure 6.3 shows that 'poor performance of equipment' was considered to be no more than 'sometimes important' by all except firm 2, which had emphasised the importance of continuing production. Interviewees gave no specific examples of the inadequate performance of energy efficient technologies and stated instead that there were often corresponding benefits such as quieter compressors. More serious was staff retraining, which was considered often important by three firms. Here again we see the ascending trend from left to right.

Figure 6.4 shows that 'the difficulty/cost of obtaining information on the energy consumption of purchased equipment' was considered sometimes important for four firms. Lack of time, however, was a serious consideration, with most firms grading this as often important, and only firm 1 assessing it as rarely important. The managing director in firm 1 strongly believed that energy efficiency saved money, so this initial belief would help to make energy efficiency a priority for this firm.

Table 6.9 summarises the results by grading each category of hidden cost according to its perceived importance. It can be seen that the absolute differences in the scores are relatively small. Both 'lack of time/other priorities' (first) and 'identifying opportunities, analysing cost effectiveness and tendering' (joint second) may be considered as indirect measures of the staff overhead costs associated with implementing energy efficiency opportunities. Hence, one conclusion is that these overhead costs are considered to be very significant. A second is that a high priority is given to avoiding disruptions to production.

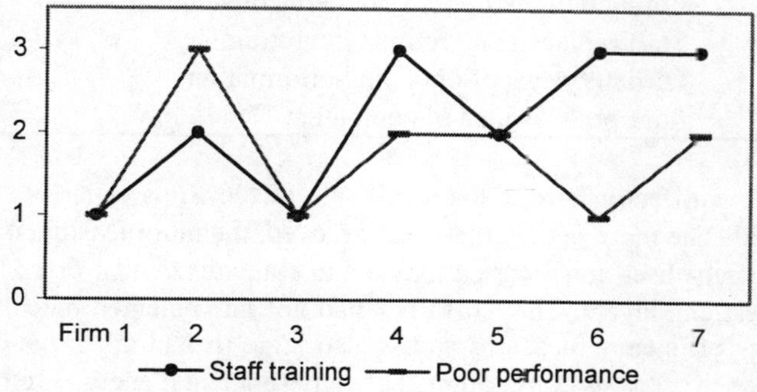

Note: On the vertical axis, 1 = rarely important, 2 = sometimes important, 3 = often important.

Figure 6.3 Hidden costs – importance of staff retraining and possible poor performance of equipment

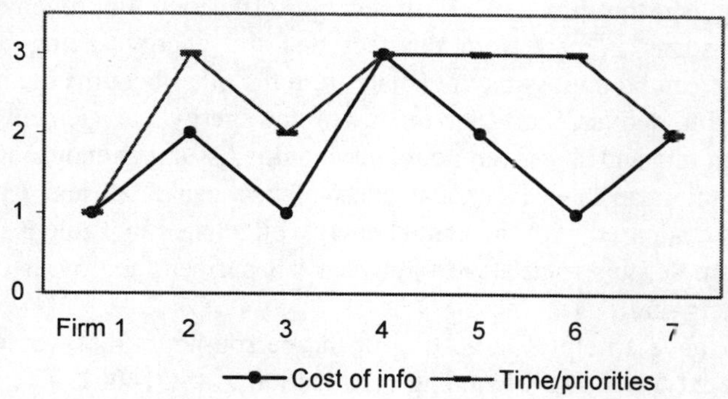

Note: On the vertical axis, 1 = rarely important, 2 = sometimes important, 3 = often important.

Figure 6.4 Hidden costs – difficulty/cost of obtaining information on equipment and lack of time/priorities

Table 6.9 Ranking of hidden costs by importance

Rank	Hidden cost	Score
First	Lack of time/other priorities	17
Equal second	Identifying opportunities, analysing cost effectiveness and tendering	16
	Production disruptions/hassle/inconvenience	16
Fourth	Staff replacement, retirement, retraining	15
Equal fifth	Difficulty/cost of obtaining information	12
	Poor performance of equipment	12

A relevant factor here is the small size of the firms in terms of numbers employed. The more people that are employed, the more feasible it is for staff to have specialised functions. It may be no coincidence that firms 1 and 2 are the largest employers. Though firm 7 had not implemented much in the way of energy efficiency measures and is also large, it had taken the decision to undertake a serious programme of environmental/energy reforms. Our findings are consistent with other studies, such as Gruber and Brand (1991), who show that SMEs are disadvantaged since they lack individuals who can be 'afforded' time to develop specialised energy management skills.

Having identified the significant hidden costs, it would be helpful to estimate their magnitude. How much staff time is devoted to 'identifying, analysing, tendering' and the like? Unfortunately most firms could only estimate the hours spent on day-to-day energy management, and were unable to distinguish the time spent on energy efficiency improvements. This difficulty stems in part from the fact that the energy saving part of an investment can be impossible to isolate from the other benefits the investment is intended to deliver. Staff time on day-to-day energy management amounted to between one and seventeen hours per month. Assuming annual salary costs of £33k, this translates to annual costs of between £228 and £3876. This should be compared with the annual energy bills listed in Table 6.7. For firm 4, for example, this translates to between 0.3 per cent and 5 per cent of the annual energy bill.

Firm 7 gave a helpful indication of the person-hours that would be spent over the next few years on gaining certification to ISO14001. The operations manager charged with the implementation of this 'for my sins' (vividly indicating a perception of hassle and inconvenience) estimated that it would involve 4–5 people spending ~50 hours per week. This could amount to staff costs of £40k–50k/year and would include the personnel time devoted to environment, health and safety as well as energy efficiency.

Taking a (high) estimate that 20 per cent of this relates to improved energy management, this leads to staff costs of £8k–10k/year, which needs to be seen in the context of an annual energy bill of some £280k. Working backwards by applying their three year payback rule and assuming for the sake of argument that a 10 per cent saving on their annual energy bills (£28k) is offered by advanced technology, they could justify an investment of £84k, ignoring discounting. If they intend *in any case* to invest in cleaner, safer technology and if there is a choice of equipment that varies only by energy usage, this is the extra amount that they could spend on the energy efficiency component. The personnel costs of £8k–£10k could be seen to represent a sizeable addition or share of the investment.

In practice, attributes of technologies are not usually separable in this way and it is not easy to isolate investment costs attributable to energy efficiency measures, let alone hidden costs. Therefore, even with a real example, it is difficult to calculate the hidden costs ratio.

4.3 Imperfect Information

Information on the size and pattern of energy use was generally poor in the case study firms and energy information systems were relatively crude. Two firms had undertaken a full energy audit that pointed to good opportunities that were subsequently taken up, while a third considered that their audit merely told them what they already knew. The Electricity Supply Board had audited compressor usage in one firm which led them to shift the timing of electricity consumption to avail themselves of better rates. In general, where the audit was adequate it did help to overcome information barriers.

Two firms considered that they had access to good information on energy efficiency opportunities, derived primarily from the Irish Energy Centre, while others felt they lacked information. Data on the energy consumption of equipment and of new and refurbished buildings was considered difficult to obtain, but generally the problem was not so much the availability of information but the lack of time to use it.

Figure 6.5 shows in summary form the usefulness of the following four information sources to the case study firms: a) the Irish Energy Centre (denoted IEC); b) professional associations, trade and technical journals (denoted Assoc/jls); c) technical conferences and seminars; and d) energy suppliers. Firms were asked to grade these sources as either excellent (4), good (3), average (2), poor (1) or don't use (0).

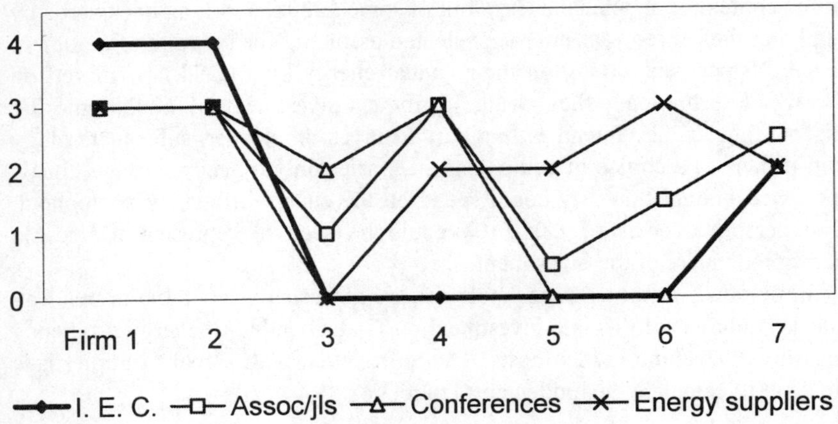

Note: On the vertical axis, 0 = don't use, 1 = poor, 2 = average, 3 = good, 4 = excellent.

Figure 6.5 Firms' ranking of the usefulness of information sources

The two firms at the left-hand side of the figure are those that had implemented the most measures and these are the ones that declared the Irish Energy Centre to be an excellent source of information. Firm 7 also found the Irish Energy Centre to be useful, but the others did not use it. Professional associations and trade and technical journals (for example, the *Plant Engineer*) were considered to be good, as were technical conferences and seminars. Energy suppliers appeared to replace the other sources for those firms that had implemented the least measures. Other information sources evaluated by the respondents included equipment suppliers (poor) and professional networks (good to poor). The only 'excellent' verdicts were awarded to the Irish Energy Centre (two firms) and 'colleagues within the company' (two firms).

Firms were also asked about their awareness of grant schemes and government information programmes. Several of these have been funded by the Irish government since the early 1990s (Box 6.2), but the level of awareness among the case study firms appeared to be poor. Several firms were not aware of any grant schemes while another heard about a scheme too late. Another claimed, incorrectly, that their firm was small and therefore not targeted while a fourth was aware of the existence of the Irish

Box 6.2 Irish policies to encourage industrial energy efficiency

The *Operational Programme for Economic Infrastructure 1994-1999* included a sub-programme on Energy Efficiency, with a funding of £34 million over the programme period. Co-ordination and implementation of the energy efficiency programme were the responsibility of the Irish Energy Centre (IEC). The principal elements of the programme that were relevant to manufacturing industry were:

- *Energy Audit Grant Scheme (EAGS):* This provided grants to organisations that engaged consultants to carry out energy audits and surveys. A grant of 40 per cent could be awarded, subject to a maximum of £5000.
- *Energy Efficiency Investment Support Scheme (EEISS):* This provided grants to organisations that invested in energy-saving technologies or measures. A grant of up to 40 per cent of the cost could be awarded, subject to a maximum of £100 000 per site.
- *Energy Self-Audit and Statement of Energy Accounts Scheme*: This provided a formal framework within which an organisation could make energy a strategic component of corporate policy. Companies in the scheme made a public commitment to assess energy consumption, define targets and strategies for reduction, and provide an annual statement on energy performance. The scheme was directed primarily at energy-intensive industry. While there was a relatively small number of participants, these accounted for over 25 per cent of industrial expenditure on energy.
- *Best Practice Programme*: This aimed to promote energy efficient technologies and practices. Information on best-practice technologies was transmitted through seminars, workshops, site visits and publication of best-practice guides and case histories.
- *Steam System Boiler Evaluation Scheme*: This aimed to encourage fuel savings in the use of boilers, through grant aid, training and awareness programmes, and a boiler awards scheme that publicly acknowledged good practice.

These schemes were applicable to the mechanical engineering industry, among many others. However, since mechanical engineering is not energy-intensive, it was not considered a priority sector.

Source: Greer (1997)

Energy Centre but was already overwhelmed with information ('you would go brain dead if you tried to take it all in').

In the context of severe time constraints, information must be both accessible and easily digestible if it is to be of value. Information also needs to be well targeted, concise, and relevant to the firm's specific circumstances – including the technical ability of the relevant staff. The case study firms preferred information that was targeted at their specific industry, relevant to their process technologies and informative about new equipment. Of particular value were case studies that included 'before and after' figures of the energy consumption of specific plant. The preferred method of receiving information was half-day conferences or seminars. Emphasis on the word 'relevant' suggested a gap between available information and that required to implement practical on-site solutions.

In sum, lack of information appears to be an important barrier for the case study firms and much of the information that is available is not geared closely enough to their specific requirements. Having an energy audit is possibly the most useful measure that could be supported

4.4 Other Barriers

Risk and split incentive problems appeared to be less important for the case study firms. These are briefly discussed below, together with two other issues that appeared relevant: the personal incentives for energy saving, and the role of environmental values within organisational culture.

Risk

Business risk could lead to more cautious behaviour and delay or reduce investment in non-essential measures, while technical risk could inhibit investment in specific energy efficient technologies. In the replies, technical risk was considered no more than sometimes important and only firm 6 thought that market uncertainty was important (Figure 6.6).

In subsequent elaboration of the kinds of risks facing them, firms mentioned currency fluctuations, a large customer in financial difficulty, reduction in product demand, increased international competition, the rapidity of technical change and the tendency for process machinery to become obsolete. These risks could potentially justify the use of strict investment criteria in some circumstances, but this rationale was never explicit. In general, the evidence for technical or business risk contributing to the neglect of energy efficiency opportunities appears limited.

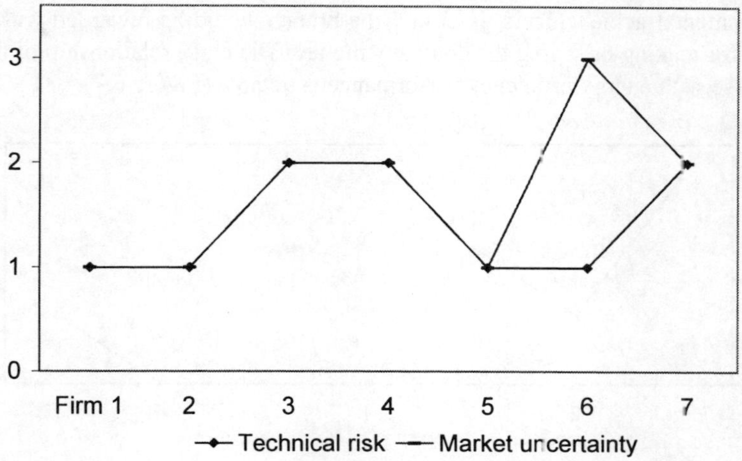

Note: On the vertical axis, 1 = rarely important, 2 = sometimes important, 3 = often important.

Figure 6.6 Importance of risk as a barrier to energy efficiency investment

Split incentives

Split incentives prevail when those implementing energy efficiency measures are not beneficiaries of their efforts. The only area where this seemed relevant for the case study firms was the accountability of individual departments for their energy costs. Several interviewees considered that departments would change their housekeeping behaviour if they were made more accountable, but they thought it unlikely that large savings would be made. Problems related to the cost of sub-metering, the small size of the incentive compared to other budgetary pressures and the limited expertise of departmental staff in managing energy consumption. Hence, split incentives did not appear to be a major issue for the case study firms, though it is interesting that interviewees strongly supported the principle of departmental accountability.

Personal incentives

The issue here is the risks and rewards for individual staff members who engage in energy efficiency improvements. While a reward system is possible in principle, there are difficulties and costs associated with monitoring the performance of individual improvements. While the energy information systems in the case study firms did not allow the performance of such measures to be adequately assessed, four firms nevertheless used some form of reward system. In firm 1 the reward was quite substantial, in firm 2 the respondent received recognition and in firm 3 there were bonuses for environmental achievements. In firm 7 the branch would be rewarded with a bonus for making or saving the company money. The close relationship of the incentive with energy efficiency performance is notable (Figure 6.7).

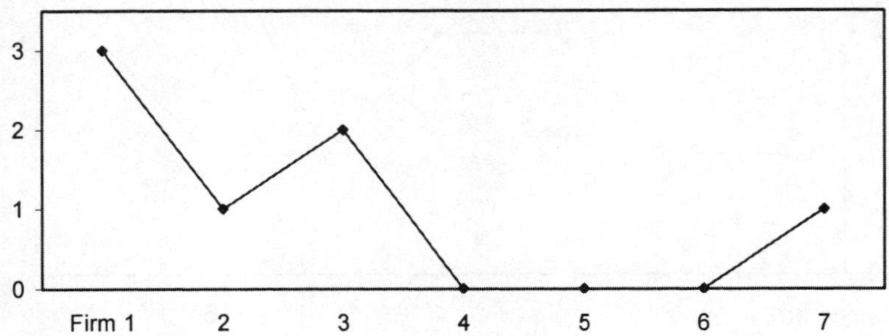

Note: On the vertical axis, 3 = substantial reward, 2 = reward, 1 = recognition, 0 = no reward.

Figure 6.7 Is the 'energy manager' or equivalent rewarded for energy efficiency improvements

Values and organisational culture

In addition to the financial incentive for improving energy efficiency, individuals and firms may be motivated by a desire to improve the firm's environmental performance. In some cases, this may have (or be expected to have) an influence on shareholder value.

Firms 3, 4, 5 and 6 did not have a culture strongly motivated by environmental values, either on the part of respondents or of senior management. Respondents from the other three firms that had or were about to implement significant energy efficiency measures exhibited a higher level of environmental concern and emphasised the importance of management commitment. In firm 1 the MD raised the profile of energy in the organisation, though emphatically for cost-saving reasons, while in firms 3 and 7 it was the head office that was pushing the environmental agenda. In one firm in particular the new emphasis on the environment derived from the importance of image and its potential effect on share price. However, 'energy is not seen to be as important as the environment', apparently, so that effort tends to be directed more towards broad-based environmental improvements rather than energy efficiency.

4.5 Summary of Barriers

Respondents were asked to sum up by saying what they thought was the major barrier to energy efficiency. Most said it was shortage of time. The exceptions were one firm that said that they had exhausted all opportunities, even if a 10-year payback were to be applied, while a second felt that the major barrier was access to capital. Despite the limitations of case study research, the insights are useful and coherent and there is a good degree of consistency within and between the replies. Table 6.10 summarises the results from the preceding sections and shows the relative importance of each of the postulated barriers.

Some of these barriers may be unavoidable, such as the staff costs associated with identifying, evaluating and implementing energy efficiency investments. However, while some transaction costs are unavoidable, it is possible that others may be reduced through public policy or changes in organisational routines. The next section explores the possibilities.

5. POLICY IMPLICATIONS

There is scope for individual firms to improve energy management, although staff time constraints create a bigger obstacle in mechanical engineering than in either brewing or higher education. Recommendations are widely available in the best-practice literature and have much in common with those reviewed in earlier chapters. They include: better use of information sources such as the Irish Energy Centre (now Sustainable Energy Ireland); conduct of an energy audit; installation of sub-metering during refurbishment or when investing in major items of equipment; replacement of simple payback calculations with NPV; allocating energy management responsibilities to a single member of staff with clear areas of control; and introducing a system of rewards and recognition. Since these initiatives require direction from the top, the biggest challenge is to encourage senior management to put energy efficiency 'on the map'. This is the role of public policy, which is discussed below at two levels: sector-specific and national.

5.1 Sector-Specific Policy

Most firms in the mechanical engineering sector are SMEs which typically lack expertise in energy matters. It is difficult to inform and motivate such firms, especially when energy prices are low. Sector-specific information programmes are likely to be inappropriate as the sector lacks distinctive characteristics. Instead, there is a need to market cross-sectoral publications

Table 6.10 Summary of the importance of barriers to energy efficiency in the Irish mechanical engineering sector

Potential barrier	Assessed importance (respondents' average scores in brackets)	Comments
Access to capital	High (2.4)	Access to borrowing is not a barrier, but strict rules on investment criteria and capital budgets may prevent some good investments. Rarely use NPV.
of which:		
Other priorities	(2.6)	
Lack of capital	(2.4)	
Strict adherence to budgets	(2.3)	
Hidden costs	High (2)	Management time is the most serious hidden cost, particularly for the smaller firms.
of which:		
Time/other priorities	(2.4)	
Analysing/tendering	(2.3)	
Disruptions/hassle	(2.3)	
Staff training	(2.1)	
Poor performance of equipment	(1.7)	
Difficulty/cost of obtaining information on equipment:	(1.7)	
Imperfect information	High (1.9)	Information on use of energy by equipment and buildings is poor. There is scope for more energy audits and greater use of the Irish Energy Centre.
of which:		
Opportunities for energy saving	(2)	
Energy use of equipment	(1.7)	

Risk	Medium (1.6)	Genuine risks exist which may contribute to strict payback rules, but evidence for this is limited.
of which:		
Market uncertainty	(1.7)	
Technical risk	(1.4)	
Split incentives	Medium to low	Considered important in principle but may be difficult, costly or impractical to overcome.
of which:		
Accountability of departments	(2.1)	
Building/equipment leasing	low	
Job rotation	low	

Note: The scores in brackets are respondents' scores, where 1 = rarely important, 2 = sometimes important, 3 = often important.

such as those on space heating, compressed air, motors and drives, heat treatment and industrial buildings. The material needs to be straightforward, easy to understand and relevant to the needs of SMEs. Improved information programmes should follow from the increased allocation of funds to Sustainable Energy Ireland in accordance with the *National Development Plan 2000–2006* (Stationery Office, 1999) and the government's *Green Paper on Sustainable Energy* (DPE, 1999). It is intended that activities under the new programme will concentrate less on grants and subsidies and more on provision of information, education, stimulation of the energy efficiency market and on research and development.

One way to add value to information programmes would be to form partnerships with relevant trade associations. Unfortunately, mechanical engineering has a diverse range of trade associations which historically have taken very little interest in energy efficiency. Alternative marketing routes could include professional bodies, specialised journals and half-day seminars. One interesting possibility is to work with large, influential customers of capital goods who could promote energy efficiency directly through their supply chains.

Benchmarks are an invaluable tool for energy management, but there is a lack of adequate benchmarks for the mechanical engineering sector. This is partly a consequence of the heterogeneity of the sector, but it also reflects a lack of analysis of energy-use patterns. There appears to be scope, therefore, for survey and research work to improve the data on energy use and to develop preliminary benchmarks. A feasible goal would be benchmarks on fossil fuel use per unit floor area for different industrial buildings, since this is predominantly for space heating. Ready reckoners and software tools for calculating performance, combined with support services, may also be valuable.

In brewing and higher education, environmental management systems (EMS) appeared to be largely irrelevant to the promotion of energy efficiency (chapters 3, 4 and 5). The situation is different in mechanical engineering, however: first the sector starts from a much lower basis in terms of policies, procedures, resources, and the level of technology adoption; and second, the alternative drivers for energy efficiency are much weaker in this sector, making it harder to get energy efficiency on the strategic agenda of management. As a result, improved environmental management may benefit energy efficiency, even if energy is seen as a secondary environmental issue. The larger companies in the sector, and particularly the multinationals, may adopt EMS out of concern for a 'greener' corporate image, but this process has been relatively slow to date. Since the sector does not have a direct interface with the public, corporate image is less of an issue. What may be of greater importance is supply chain pressure from large customers such as vehicle manufacturers, who are establishing preferred supplier systems and supply chain initiatives. These involve a move away from competitive tendering towards partnering and long-term relationships. Here the emphasis is on improving quality and service, rather than competing narrowly on price (see chapter 7). Since many of the largest companies are concerned about their environmental profile, there has been a natural extension of supply chain initiatives to include environmental issues. Typically suppliers are encouraged or required to accredit to ISO14001.

These are important developments since they affect many smaller firms which would not normally consider environmental management. The process may be encouraged by government through conducting surveys and publishing data on the take-up of EMS in different sectors; evaluating the quality of information and support services for EMS, with the aim of identifying areas for improvement; providing guidance on energy efficiency to EMS consultants and certification bodies; including information on energy efficiency (and pointers to information) in broader publications on environmental management; highlighting the cost-saving benefits of EMS through case study publications – with particular focus on energy efficiency; and developing guidance on EMS, targeted more specifically at the engineering sector.

5.2 Energy Efficiency Policy at the National Level

The introduction of carbon taxes and emissions trading should increase energy prices, attract management attention and make some of the identified barriers relatively less significant. However, it is unlikely that such measures will transform energy management in the sector since energy will remain a very small proportion of total costs.

Proposals for a carbon tax, and for emissions trading in the case of large emitters, were included in the Irish government's *National Climate Change Strategy* (DELG, 2000) and both of these policies are to be applied. A gradual switch from labour taxes to carbon taxes or auctioned permits, with due regard for implementation issues and political sensitivities, could on balance benefit the Irish economy (Bergin et al., 2004; McCoy and Scott, 2002; Scott, 1999). In particular a switch in taxation away from labour taxes to a carbon tax could improve the profitability of the mechanical engineering sector, since it is labour-intensive rather than energy-intensive. Firms will see the cost of emitting carbon rise and the cost of labour drop. Appraisals of energy efficiency investments will become more profitable and the 'hidden' salary costs will be relatively less important.

To enhance the effect of a carbon tax, one possibility is to use a portion of the revenues to fund specific initiatives to improve energy efficiency. These could include demonstration schemes, support for energy efficiency training, subsidised energy audits, on-site consultancy support and either subsidies or tax allowances for investment in specific types of energy efficient technology – such as condensing boilers, energy efficient lighting, variable speed drives and insulation. The last proposal directly addresses the access to capital barrier and could be more effective than either energy price increases or information programmes acting alone – although the scheme would need to be large to reach more than a fraction of eligible firms. An effective way of multiplying the impact of such schemes would be to establish qualifying criteria that are not linked to the investment itself. For example, subsidies or tax allowances might only be given to firms that had conducted an energy audit and had committed to undertaking the identified measures. In this way, the measure would encourage the take-up of technologies over and above the specific qualifying investment.

One drawback with subsidies and tax allowances is that they can entail excessive administrative costs, both for government and the target firms. As pointed out by Gruber and Brand (1991), small firms rarely have time to read brochures or to fill in application forms. Subsidies need therefore to be straightforward and well-publicised. In the above-mentioned study, some 57 per cent of SMEs in West Germany had not heard of a subsidy programme on energy conservation that had existed for ten years.

Another measure that could facilitate investment is third-party finance, including facilities management and energy service contracting. ESCO-type operations are in the early stages in Ireland and focus mainly on large-scale, energy supply installations such as CHP plants within large, energy intensive firms (Sorrell et al., 2000). The difficulty for the mechanical engineering sector is that the energy bills of most firms are too small for energy service contracting to be viable. While multiple site contracts may be an option, relatively few firms in this sector have more than one site. Hence, the best hope is for innovations such as standard-form contracts to reduce the viable size threshold.

While the above initiatives may help to improve energy management within individual firms, there are severe limits to what can be achieved through this approach. Energy is likely to remain a marginal concern for most firms, while staff time constraints will continue to inhibit energy efficiency improvements. This is particularly the case for new equipment purchases where energy consumption is normally overlooked. An alternative, therefore, is to simplify organisational decision-making (that is, reduce transaction costs) by *transforming the market* for certain types of energy-using technologies. The options are to introduce energy labelling of goods, impose mandatory energy efficiency standards, or encourage improvements in energy efficiency through voluntary agreements with manufacturers. The requirements of the internal market mean that such initiatives generally need to be taken at the EU level.

Market transformation programmes are well established within the EU, particularly with regard to domestic appliances, and there is considerable scope for expanding these initiatives into the non-domestic sector to include technologies such as electric motors, lighting and refrigeration. In 1999, an agreement between the European motor manufactures association (CMEP) and the Commission resulted in classification of energy efficient motors into three categories, together with a commitment to halve sales of the least efficient category from 2003 onwards (European Commission, 1999). With its high use of machine tools, the CMEP agreement should have a significant impact in the mechanical engineering sector, since motors and drives account for up to 60 per cent of electricity use (a proportion which is increasing). Since most of these motors are embedded in machine tools and other equipment, the scope for the end-user to influence motor choice is relatively limited. Market transformation initiatives are therefore essential to improve energy efficiency in this and related areas. This type of approach should effectively circumvent organisational barriers to energy efficiency.

6. SUMMARY AND CONCLUSIONS

This chapter has described the investigations into barriers to energy efficiency in the Irish mechanical engineering sector. It was shown that the dominance of SMEs, the concentration of energy use in generic process and building applications, the small contribution of energy to total costs and the severe constraints on staff time all contributed to a general neglect of energy efficiency.

Seven firms were selected and agreed to participate as case studies, representing 13 and 14 per cent respectively of the sector's employment and sales. For ease of discussion, the firms were ranked according to the number of energy efficiency measures that they had implemented. This ranking revealed that the firms that had adopted most energy efficiency measures tended to be the larger firms with the larger energy bills. These had conducted energy audits, established reasonable energy information systems and perceived there to be fewer unexploited opportunities for energy efficiency improvements. The opposite was the case for the other firms.

The most important barrier to energy efficiency was found to be limited access to capital, with energy efficiency investments being given a low priority in comparison to strategic and non-discretionary investment. A second and related factor was hidden costs and in particular the staff time required to identify, evaluate and implement investments. Smaller sites had particular problems, since it is difficult to allocate a fraction of an individual's time to energy management, while other priorities repeatedly received precedence. The final, and also related factor, was the lack of information on the opportunities for energy efficiency improvements. This could be linked to constraints on staff time, lack of relevant skills, inadequate sub-metering, the high transaction costs entailed in identifying the energy performance of new equipment and the general neglect of energy as an issue. Also, the available information sources appeared to be of poor quality and lacking in relevance

Evidently a range of policy measures is required to address these barriers. A selection of possible measures was proposed at the firm, sectoral, and national level which are interlinked – individual firms are unlikely to take initiatives until energy is given a higher priority by senior management and this in turn is unlikely to occur without higher energy prices. Carbon pricing is a necessary but not sufficient element of the policy mix. A revenue neutral tax is likely to be of benefit to the mechanical engineering sector, while the 'stick' of energy taxation can be supported by 'carrots' such as subsidised energy audits. Broad-based encouragement of environmental management schemes may also be of benefit in this sector.

The aim of policy should be to provide accessible, relevant and credible information on energy efficiency opportunities and to encourage energy audits, energy monitoring, improved energy labelling, energy performance benchmarking, more informative energy bills and greater energy numeracy. This should occur in a supportive regulatory and pricing environment that is as risk-free as possible and accounts for external costs. But there are limits to what can be achieved by individual energy-using organisations acting alone. To a greater extent than in the brewing and higher education sectors, energy efficiency improvements in this sector must rely on actions taken further up the energy service supply chain. This includes minimum efficiency standards and market transformation programmes for key items of equipment such as electric motors. Multiple barriers require multiple responses. But the guiding theme should be to make it easier for organisations to make choices that improve energy and economic efficiency.

NOTES

1. The full results from all three of the Irish case study sectors are published in O'Malley et al. (2003).
2. Case study interview, spring 2000.
3. The approximate breakdown of products sold to different sectors is: 70 per cent to manufacturing industry; 4 per cent each to mining, energy and water supply, agriculture and forestry, and the building sector; and 14 per cent to the service sector (European Commission, 1996, pp. 13-15). Within manufacturing, the industry itself is its own biggest customer, followed by the automobile and electrical engineering industries.

4. All monetary values in this chapter are expressed in Irish pounds (£), where one Irish pound equals 1.27 euros.
5. These figures refer to NACE Revision 1, Sector 29, 'manufacture of machinery and equipment not elsewhere classified (n.e.c.)', excluding category 297, 'manufacture of domestic appliances'. The remaining statistics quoted in this section refer to the whole of sector 29, without excluding 297, because of data constraints.
6. Approximate figures are given in O'Malley et al. (2003).

REFERENCES

Barry, F., J. Bradley and E. O'Malley (1999), 'Indigenous and foreign industry: characteristics and performance' in F. Barry (ed.), *Understanding Ireland's Economic Growth*, MacMillan, London.

Bergin, A., J. FitzGerald and I. Kearney (2004), *The Macro-economic Effects of Using Fiscal Instruments to Reduce Greenhouse Gas Emissions*, Irish Environmental Protection Agency, Johnstown Castle, Co. Wexford.

CSO (Central Statistics Office) (2000), *Census of Industrial Production 1997*, Central Statistics Office, Dublin.

DELG (Department of the Environment and Local Government) 2000, *National Climate Change Strategy, Ireland*, Stationery Office, Dublin.

DPE (Department of Public Enterprise) (1999), *Green Paper on Sustainable Energy*, Stationery Office, Dublin.

Energy Efficiency Office (1993), 'Fuel efficiency booklet no. 1: energy audits for industry', ETSU, Harwell.

European Commission (1996), *Panorama of EU Industry 1995–96*, Directorate General III, Brussels.

European Commission (1999), *Actions to Promote Energy Efficient Electric Motors*, Motors Study Group, Directorate General for Energy, October.

Greer, H. (1997), *Energy Efficiency Policies and Programmes in Ireland: A Review*, Network Resources Limited, Dublin.

Gruber, E. and M. Brand (1991), 'Promoting energy conservation in small and medium-sized companies', *Energy Policy*, **19**(3), 279–87.

McCoy, D. and S. Scott (eds) (2002), *Green and bear it Implementing Market-Based Policies for Ireland's Environment*, proceedings of a conference held on 10 May, Dublin: Economic and Social Research Institute.

O'Malley, E., S. Scott and S. Sorrell (2003), *Barriers to Energy Efficiency from Selected Sectors*, Policy Research Series No. 47, Economic and Social Research Institute, Dublin.

Scott, S. (1999), 'Fiscal measures for a sustainable economy – Ten fears about a carbon tax', Chapter 5 in *Sustainable Energy and the Environmental Imperative,* Institute for European Affairs, Dublin.

Scott, S. (2000), 'Energy service companies in Ireland', report to DG Research on the project Barriers to Energy Efficiency in Private and Public Organisations, Economic and Social Research Institute, Dublin.

Sorrell, S., J. Schleich, S. Scott, E. O'Malley, F. Trace, U. Boede, K. Ostertag, and P. Radgen, (2000), *Barriers to Energy Efficiency in Public and Private Organisations*, SPRU – Science and Technology Policy Research, University Of Sussex, Brighton.

Stationery Office (1999), *Ireland: National Development Plan 2000-2006*, The Stationery Office, Dublin.

7. Regulations Are Not Enough: Barriers to Energy Efficiency in the UK Construction Industry[1]

Steve Sorrell

'Most US buildings of the past few decades are monuments to the designer's ignorance of where the sun is.' (Lovins, 1992, p. 11)

1. INTRODUCTION

The energy efficiency of new buildings is critical to a sustainable future. The choices made on building form, fabric and orientation will determine a building's performance throughout its life, while the choices made on building services may be difficult and expensive to modify subsequently. New construction provides the greatest opportunity for the adoption of energy efficient technology, and the opportunities are significantly more cost-effective than in retrofit situations.

In 1994, buildings accounted for 46.6 per cent of UK final energy consumption and 46.9 per cent of CO_2 emissions (Table 7.1).[2] Non-domestic buildings in the public, commercial and industrial sectors accounted for 40 per cent of building CO_2 emissions, and 19 per cent of the UK total.[3] These proportions are comparable to other industrialised countries and suggest that non-domestic buildings should be a priority sector for climate policy. In practice, however, the sector is seriously neglected, despite rapidly increasing energy consumption (Scrase, 2000).[4] While the slow rate of stock turnover (~2 per cent/year) makes retrofit and refurbishment important, it also means that new buildings should be constructed to a high standard since many will still be in use in 50 years time. If they are inefficient, they will present a major obstacle to achieving deep cuts in carbon emissions.[5]

Table 7.1 The contribution of buildings to UK final energy consumption and CO_2 emissions

Sector	Final energy use		Carbon emissions	
	PJ	% total	MtC	% total
Commercial and public buildings	880	13.1	21.2	15.1
Industrial buildings	282	4.2	5.6	4.0
Total non-domestic buildings	***1162***	***17.4***	***26.8***	***19.0***
Domestic buildings	1960	29.3	39.2	27.8
Total buildings	***3122***	***46.6***	***66.0***	***46.9***
Industrial processes	1231	18.4	27.7	19.7
Transport	2294	34.3	46.0	32.7
Agriculture	49	0.7	1.1	0.8
TOTAL	**6696**	**100.0**	**140.8**	**100.0**

Source: Building Research Establishment, private communication.

The energy efficiency of new non-domestic buildings was a dominant theme in the UK higher education case studies (Chapter 4). All the case study institutions had major capital programmes underway and all interviewees recognised that new construction represented a major opportunity to improve energy efficiency. Despite this, each case study provided a series of examples of cost-effective energy efficiency opportunities being lost. The procurement process within individual universities played a role here, including in particular the rigidities in capital budgeting procedures (see Chapter 4). But the primary source of the problems appeared to be the structure and operation of the construction industry and the associated mechanisms for project financing. In particular, the problems derived from the incentives placed on different actors within the construction process.

The construction of non-domestic buildings provides an important example of the drawbacks of focusing solely on the end-user organisation when identifying barriers to energy efficiency. An adequate understanding of the problem requires an analysis of the full supply chain, including the incentives and constraints on all the relevant actors. Similarly, an effective policy response must also address each aspect of the supply chain. This chapter therefore departs from the framework of the previous chapters by exploring the barriers to energy efficiency within the supply chain for new non-domestic buildings. The analysis is based primarily on the results of the six UK higher education case studies, supplemented where necessary by additional documentary research and interviews with construction industry specialists. But while the empirical basis is relatively limited, the results are consistent with the available literature on this topic, and in particular with

Amory Lovins's excellent analysis of institutional barriers to energy efficient buildings in the US (Lovins, 1992).[6] While the themes have long been familiar to construction industry specialists (Bowley, 1966), they have received relatively little attention within the academic literature on energy and climate policy. This is an important omission.

The UK construction industry is currently undergoing major reforms, stimulated by technical innovation (Gann, 2000), changes in project funding and procurement (notably public-private partnerships) (Grout, 1997), and policy initiatives based on the reform agenda proposed by two important government reports: *Constructing the Team*, or the 'Latham Report' (Latham, 1994); and *Rethinking Construction*, or the 'Egan Report' (Egan, 1998, Blockley and Godfrey, 2000). The problems discussed below are recognised within the industry and best practice in the sector is overcoming them to some extent. But the case study results suggest that: first, there is a wide gap between best and typical practice; second, the problems identified are still characteristic of the majority of new construction in the institutions analysed (and by implication the UK public sector as a whole); and third, many of these problems are found in 'green building' projects where the parties involved have made serious attempts to achieve good practice in energy performance.

The chapter first summarises some features of the UK construction industry and then highlights the most important barriers to energy efficiency identified in the case studies. These are interpreted in terms of the theoretical framework developed in Chapter 2, and a case is made for integrating climate policy objectives into the ongoing reform of the construction industry. The primary focus throughout is building services (heating, ventilation, air conditioning and lighting) as this is where some of the greatest problems appear to lie.

2. BARRIERS TO ENERGY EFFICIENCY IN THE UK CONSTRUCTION INDUSTRY

2.1 Features of the UK Construction Industry

As documented in the Egan report and elsewhere, the UK construction industry compares badly with other countries in terms of capital cost, product quality and client satisfaction (Egan, 1998; Latham, 1994). The industry is characterised by low and unreliable profitability, limited investment in R&D and persistent skill shortages.

A notable feature of the industry is fragmentation, with over 160 000 individual firms (Egan, 1998, p. 11). Construction projects are complex and require the involvement of a large number of actors – for example, client, architect, lawyer, consultant, equipment supplier, construction worker, building commissioner, and so on. Typically, construction projects are achieved by temporary coalitions of firms who relate through subcontracting. The contractual structure depends on the procurement route, but increasing technical complexity and the desire for risk transfer has led to main contractors preferring to use subcontracting rather than bear the risk of employing staff directly (Gann, 2000, p. 167). The resulting contracts are detailed and complex, and low trust, adversarial relationships between the different actors are the norm (Winch, 2000, p. 144). As the Egan report notes:

> '...the efficiency of project delivery is presently constrained by the largely separated processes through which they are generally planned, designed and constructed. These processes reflect the fragmented structure of the industry and sustain a contractual and confrontational culture...This process may well minimise the risk to constructors by defining precisely, through specifications and contracts, what the next company in the process will do. Unfortunately, it is less clear that this strategy protects the client...' (Egan, 1998, p. 22)

Typically, contractors are selected through competitive tender on the basis of lowest price, and there is little communication between the parties involved. Fierce price competition and low margins create an incentive for the winning contractors to cut costs (Winch, 2000, p. 144). This leads to the use of oversight and control mechanisms, such as quantity surveyors, to reduce the risk of opportunistic behaviour (Winch, 1989; Reve and Levitt, 1984).

The contractual culture also creates a threat of litigation. Designers are liable for defects, but some defects may be introduced in the construction process by contractors attempting to cut costs. To accommodate this, it is common to include high safety margins in designs, leading to over-specified equipment. As an example, heating or air conditioning systems may be significantly oversized in relation to the load, with the resulting part load operation being detrimental to energy efficiency.

Both these features apply in particular to building services and have important consequences for energy efficiency. The subsequent sections elaborate these problems, using examples from the UK higher education case studies.

2.2 Project Financing and Time Constraints

The sources of finance for new construction in the UK public sector have changed significantly over the last 20 years. Public investment in infrastructure has declined since the 1970s, encouraged by political commitments to reduce taxation and the national debt. In the higher education sector, the funding councils ceased to allocate grants for new buildings in 1994 (except for certain circumstances). As a result, universities now fund building projects through a variety of sources, including borrowing, internal sources, sponsorship, donation, and other public sector grants (NAO, 1998).

The increasing use of sponsorship has meant that institutions are increasingly reactive to funding opportunities. Furthermore, there are frequently major time constraints associated with these opportunities. This can create problems:

> 'We have this laboratory building...the University put in £1m of the £15m. The rest was funded by [external funders]. It was built on a speculative basis and it went like a steam train because after being a gleam in somebody's eye...suddenly the funders said they'd got to spend it, they'd got to commit the funding by March. And so, everything was done in a tremendous rush.'[7]

> 'The funding system is stop-go and completely wilful...We've got a building at the moment which...was completed to all intents and purposes 9 months ago. We've not been able to use that building because the design-build contractor who won the contract hasn't been able to get it to work properly...The speed [of construction] was forced on us because [the external funder] had to spend £4m in that financial year. They had to commit it... Consequently, they forced us down a design-build route.'[8]

> 'Sources of funding are now a complete roller coaster.. We're no longer planning on the basis of longer term, strategic plans. We're reacting almost month-to-month.'[9]

The time constraints derive from a variety of sources, including the frequent requirement of funders to spend funds before the end of a financial year. The result, however, can be a bias against energy efficient buildings as these generally involve greater design work and hence a lengthier design process.

At all stages of the construction process, it can take longer to design, assess and implement novel and unfamiliar features than to use tried and trusted solutions. To the extent that the former are more energy efficient, this creates a bias against energy efficient buildings. For example, it may be easier to design a relatively crude building and to use air conditioning to achieve adequate environmental conditions than to spend time and effort on passive ventilation or mixed mode designs – even though the latter may have lower capital and operating costs (Cooper, 1998; Bordass, 2001).

A 1996 study on over-specification in building services commented on the lack of time that contractors and suppliers had to prepare bids (Gann et al., 1996, p. 25). This, combined with the low profit margins and lack of trust, frequently resulted in inappropriately specified bids. There was a widespread practice of using bids from previous work and modifying them only slightly to meet the current contract. The result was outdated specifications, with no real consideration of the client's needs. Exactly this type of problem was encountered in the case studies:

> '...it can be that a Rolls Royce will be specified out of force of habit because that's always the way they've done it. We've got examples where the mechanical services consultant has proposed "boilerplate" control specifications which are built to some incredibly high specification which is satisfactory for working the North Sea oil installations – because they just pasted it over from another contract document. So then as a consequence, it's over-priced, over-specified and inappropriate.'[10]

Capital constraints are commonly seen as the reason efficiency opportunities are missed. But a closer investigation suggests that time constraints can be equally important:

> 'This time imperative was the primary reason for opportunities being missed. Not getting the right people together to talk about ideas. Not leaving space to be creative...Instead, the ethos was we need to know now...At the beginning of the process all the right people were involved. They got relevant top academics, financial consultants, architects, etc. They got some great ideas. But after that it was all lost. Consultants took over. They got the building up in time and only a little over budget. But most of the environmental features were dropped...'[11]

> 'Contractors and staff all wanted more time...They avoided new initiatives as they can drain time...You find that however much information you get about new technologies, there will always be something unexpected that you have to deal with. This takes time. The funding process militates against this. It encourages caution and conservatism. You do what is safe.'[12]

As this example shows, time constraints pervade the construction process and are not confined to project funding. This was evident in a number of the case studies, where the desire to complete projects before the beginning of an academic term had led to energy efficient options being dropped. Self-imposed constraints thereby exacerbate those from project financing.

2.3 Neglect of Whole-Life Costs

An inefficient building imposes a cost penalty on the client throughout its lifetime. While the client has an incentive to minimise whole-life costs, the contractors and consultants do not as they have no long-term interest in the building and are not accountable for performance in use. Increasingly, public sector guidance on construction procurement and best-practice literature is emphasising the importance of whole life costs (HM Treasury, 2000c). But the case study results suggest that such guidance is frequently not followed. The project financing arrangements mentioned above, coupled with incentives to maximise floor area for a given budget, can militate against whole life costing. All the case studies provided examples of projects where energy efficient options had been dropped in order to come within the available budget – despite having rates of return which substantially exceeded the cost of capital to the organisation.

> 'Things like low energy lighting get treated as an optional extra...These are the first to go. The building will still function without them... The project manager's main objective is to build to the budget. There is a contingency fund built up, but this is easily used up. Energy efficiency gets dropped unless there is a product champion pushing it...'[13]

This tendency appears to be reinforced by the reliance on £/m^2 capital cost guidelines for different types of building, derived mostly from precedent. The risk here is that the use of these as rigid guidelines can bias clients against energy efficient buildings, since these tend to have higher capital costs, while performing better on a whole-life basis.

2.4 Lack of Integration

The fact that green buildings require an *integrated* design team, combining a wide range of different specialists, is well established (Wilson et al., 1998; Austin and Steele, 1999). It is only through in-depth collaboration that the complexity of trade-offs between architectural features, building services and other factors can be achieved (Lovins, 1992, pp. 16–18). For example, more efficient windows may reduce lighting needs, lead to lower passive heat generation and consequently reduce air conditioning loads (Lovins and Lovins, 1997). Such approaches remain rare, however:

'The Task Force...sees the industry dealing with the project process as a series of sequential and largely separate operations undertaken by individual designers, inspectors and suppliers who have no stake in the long term success of the product and no commitment to it. Changing this culture is fundamental to increasing efficiency and quality in construction.' (Egan, 1998, p. 16)

Or as a trade article on building services put it:

'All too often elements of a building are designed in isolation, and the quality of information that passes between consultants and contractors is highly variable and open to misinterpretation. The result? A contractual quagmire and dysfunctional building services.' (Bunn and Smith, 1999, p. 29)

As a consequence, the *integration* of the construction process and team around the product was one of the core recommendations of the Egan report (Egan, 1998, p. 16). The case studies provided examples of where such integration had been successfully achieved in the higher education sector. But the results also suggest that such integration is relatively rare; indeed, it had not been achieved on other projects going ahead at the same time at the same institutions. The sequential, or linear, design process still appeared dominant:

'Green buildings are possible if an integrated design team is appointed at the beginning and correct instructions are given. But it is hard to find such teams. Architects are rarely willing to compromise their design integrity and defer to M&E[14] requirements at an early stage. There are some consulting engineering teams who are "in there" working with the architects and who evolve an integrated solution from the beginning. But both the engineers and architects who do this are very few on the ground.'[15]

'We're extending a 1960s gym building and putting...a training pool on the west side. It's such a classic because, yes, pool energy efficiency was flagged right from the beginning. We gave the design teams the Energy Efficiency Best Practice Programme material and we've ended up with complicated mechanical ventilation systems in the office accommodation which is on the perimeter. There's a sophisticated, energy efficient ventilation system in the pool, but they're putting a tin shed around it! Its roof has a 0.45U value because that's what the building standard says…The building in this case was built in a linear fashion in that the architects did the architecture, and then they gave a bit of space to the building services engineers and then they took it to Planning and had a whole lot of the roof plant room taken away...This is what I mean by linear design. And this is despite having a client who tabled environmentally friendly criteria with the design teams at the beginning.'[16]

This problem stems in part from a lack of communication and understanding between diverse professional disciplines (Lovins, 1992, p. 15). The limited knowledge of building services among architects was mentioned by a number of interviewees and relates to the professional culture of architecture and the teaching practice of architectural schools. But the problem is reinforced by the fragmented nature of the construction industry, the dominance of subcontracting, and the fact that it is common to use a new team of designers, constructors and suppliers for each project. This inhibits co-ordination and learning and prevents the development of skilled and integrated teams who may be able to deliver green buildings.

Initiatives to overcome such difficulties are central to the Egan report's recommendations. There are professional teams in the UK who adopt a multidisciplinary approach and are successful.[17] But they need clients to ask for such an approach and pay the resulting premium. Pressure to reduce design fees can militate against such an approach, despite the fact that the cheapest is not necessarily the best.

2.5 'Trimming' of Building Services

An important consequence of this lack of integration is the *sequence* of design work and the fact that building services typically come last in the sequence. Building services engineers are frequently presented with building designs – including orientation, form, layout and electrical loads – that are relatively finalised and difficult to change (Lovins, 1992, p. 18). This means that building services have to be fitted in as best as possible, rather than co-optimised with the rest of the design.

If building services come last, they are also most at risk from budget shortages or time overruns. Interview results suggest that such problems are common on construction projects where the overriding incentive is to keep to time and budget and to avoid any contractual penalties. With the complexity and novelty of major construction projects, it is common for projects to be delayed and for contingency reserves to be largely eaten up before the building services are finalised. This means that building services may suffer and energy efficient features may be dropped, despite having a good rate of return.

This problem was evident at a major new development at one university, despite serious attempts at low energy design:

'The M&E was put in after the building was constructed and it was subject to constraints on the budget since other areas of the building were overspent. There were also constraints on the engineers' time as the thing had to be up and running by the beginning of term. Time and budget are fixed. They did the best they could with the funds and time that were allocated.'[18]

'If someone like myself can get in at an early stage you can get a well engineered job. But if you're involved at the last minute you have a time constraint. You can't do much about anything. You've lost the opportunity. The gate is closed. You've got what you've got...'[19]

2.6 Over-Specification and Oversizing

One of the most pervasive factors undermining energy efficiency is the *oversizing* of heating, ventilation and air conditioning (HVAC) systems and the consequent prevalence of part load operation. For example, Brittain (1997a; 1997b) estimates that oversizing of boilers and chillers in the UK is responsible for 15 per cent or more of UK HVAC energy consumption.[20] Oversizing partly results from the lack of integration of building design noted above. However, it is compounded by the incentives on engineering designers to reduce their legal liability and to build-in flexibility to accommodate future changes in building use.

Pressures to reduce the time taken in specification and design have led to more of the detailed design work being carried out by specialist contractors (Lovins, 1992, p. 12). In the absence of co-ordinated design, these contractors may lack information on the critical elements of building performance, electrical load and passive heat generation. They will also be keen to ensure that there is sufficient space for HVAC equipment to accommodate changes in the client's or architect's requirements. As a consequence, they will have an incentive to add a safety margin on equipment size. As one interviewee noted:

'The building services engineer's equipment takes up space and there's a constant spatial battle between the service engineer and the architect...In their first go at it, they would specify a larger air handling unit than they might absolutely need...They could, and they should [leave space for extra equipment]. However, if the architect sees on the building services engineer's plans a small piece of equipment in a large room he will ask for some of that space…'[21]

Since the engineers will not pay capital or operating costs, they face no penalties if equipment is oversized. Similarly, they will not receive any rewards for keeping size to the optimum. However, they may be legally liable if the equipment proves inadequate for the current load. This problem was evident in several of the case studies:

> 'My basic qualification for an engineer is somebody who is able to double the size of everything...They're particularly risk-averse because they're keen to ensure that there are sufficient margins to meet changing requirements. In the case of many a building project, as the building is developed, new requirements are suddenly dreamt up by the client and/or the architect. Then they suddenly need more rooms to have mechanical ventilation in them. So what typically happens is they'll say "Well, I need this much capacity because in my experience I've found that if I don't ask for more I'll certainly get less."'[22]

> 'The problem is, that every engineer that I know will mark it up himself – "We need this, but we'll just hedge our bets a little bit and tell them that little bit more, just to make sure we've got enough."'[23]

As the quote illustrates, the incentive to oversize is compounded by the desire to maintain flexibility in the future uses of a building through the provision of adequate capacity. Gann et al. (1996) found that British clients emphasised this requirement more than in other countries, and they cite a university example where oversizing was recognised as a problem but was considered acceptable given research priorities and the likelihood of future changes in building use. Similarly, the need for flexibility was a dominant feature of the case studies, with interviewees emphasising the frequency of changes in use patterns and the additional cooling requirements created by expanding computer use and the introduction of new research equipment.

Including oversized equipment at the outset is not the only way to allow for future changes in building use. It is also possible to make provision – in the form of space, pipes, ductwork, and so on – to allow for additional equipment to be accommodated in the future. Much of this relates to the design of the building shell, fabric and layout rather than the detailed specification of engineering services. It can therefore only be achieved if a more integrated approach to design is adopted, with building services engineers playing a central role in the design team.

2.7 Competitive Tendering and Loss of Quality

The traditional practice in UK construction has been to select contractors on the basis of lowest price (Winch, 2000, p. 144). Contractors have to bid very hard to win contracts and then have a strong incentive to maximise their margins by cutting corners. This problem was highlighted as a major reason for lost opportunities in energy efficiency:

> 'The university ties down the spec as best they can. The contractor then goes away and sees how cheaply they can do it for. They argue that some of the spec cannot be achieved within budget...The university will take the lowest bid tendered – from a need to satisfy the auditors apart from anything else. But the moment they have got the contract, they start negotiating over details. "Oh, you didn't specify that in the tender. If you want that, it will cost you £X more" and so on...You try and make the tender as tight as possible...They try to maximise their profits.'[24]

One element of this problem is the use of 'equivalent to' requirements in equipment specification (Lovins, 1992, p. 51). There is a risk that, without comprehensive oversight, this phrase may allow contractors to substitute equipment that is cheaper, poorer quality and less energy efficient than the client requires:

> '...there's not enough oversight to be able to pull them up. On one recent project we said "Must be RA2070, low temperature range thermostatic radiator valve" and they put in a tacky domestic version.'[25]

The financial incentives for such practices are reinforced by time pressures in construction and lack of knowledge and/or information. This may be particularly important for building services, the intricacies of which may be less familiar to project managers. Oversight and control is normally achieved through a 'Clerk of Works' whose role is to establish lists of defects for non-conformance with specifications. But, as one interviewee noted:

> 'It's exceptional that an appointment is made for a *services* Clerk of Works...to oversee these exceptionally sophisticated black box things lacing through buildings now, and to ensure highly sensitive equipment is installed correctly and to specification. As a consequence, we're finding that buildings are being handed over to us...and they just patently don't work. So we're having to fault find.'[26]

The competitive tendering process was highlighted by the Latham report as a primary cause of inefficiencies in construction and the report provided recommendations on the inclusion of quality as well as cost criteria in tender evaluation (Latham, 1994). Guidance on best practice in procurement was subsequently developed by the Construction Industry Board, established in 1995.[27] A similar theme, under the term 'Value for Money', now runs throughout UK government guidance on public sector procurement (HM Treasury, 2000a). Quality, however, is more difficult to assess than price and the recommendations are challenging to implement. The 1998 Egan report was commissioned in part in recognition of the limited impact of the reforms proposed by Latham. Egan focused on more fundamental institutional changes, notably the development of *partnering* agreements (see below).

2.8 Fee Structures

A second traditional practice in UK construction is to base engineering fees on a percentage of the capital cost of the building services. This practice is considered to be a primary cause of over-specification (Gann et al., 1996, p. 27; Lovins, 1992, pp. 23–6).

Percentage of cost fees provides a direct incentive to oversize equipment since larger and more expensive equipment means higher fees. If oversized equipment is the norm, the specification will not be questioned by the client since it falls within standard and accustomed practice. Furthermore, to the extent that energy efficiency can be improved through innovative design and optimal sizing, percentage of cost fees creates a direct disincentive to improve energy efficiency as this means more work for less money (Lovins, 1992, p. 23). The fee structure therefore reinforces the incentives for oversizing resulting from lack of integration, avoidance of legal liability and the need for flexibility. It also creates a problem of cost control for the clients if construction costs rise.

In a National Audit Office (NAO) study of the management of eleven construction projects in higher education, three of the projects used percentage of cost fees, two used hourly rates and six used fixed fees (NAO, 1998, p. 39). For the sector as a whole, consultants' fees averaged 8 per cent of total project costs. The hourly rate approach reduces the direct incentive for oversizing but gives similar problems of cost control to the percentage of cost. Fixed fees overcome this, but still fail to create a direct incentive to improve energy efficiency. Innovative design and optimal sizing would still mean more work, but this time for the same amount of money.

Reform of fee structures to promote energy efficiency must ultimately aim at making a part of the fee for the whole design team contingent upon building performance *in use*. This in turn requires investment in follow-up and feedback and implies more risk and higher initial fees – although lower whole-life costs for the client.[28]

2.9 Building Commissioning

The time constraints described above can have an important effect on building commissioning as this is inevitably last in the sequence. Overruns on other parts of the construction process can squeeze the time available for commissioning. Tight time constraints may also be imposed by the client. One interviewee gave an example of a building which was handed over in January in order to be available at the beginning of the academic term. Six months later the university was still investing substantial staff time in adjusting controls which had never worked properly due to inadequate commissioning. With the increasing complexity of building services, commissioning becomes more important and requires more time:

> '...with not only energy management requirements but also a whole host of electronic sophistication in modern buildings...we reckon there needs to be a commissioning period of 8 weeks...This process gets ditched. A classic was a big Chemistry extension...they had some problems, early delays...and they lost 7 weeks before they got out of the ground. They kept assuring the client that they were going to achieve the contracted timetable, because of course they're liable for punitive damages if they run over, and they took 7 weeks out of the 8 weeks commissioning time and the rest ran over 2 weeks and so consequently there was minus 1 week remaining to commission all the services.'[29]

Alongside these time pressures can go lack of training of the staff responsible for operating the building and lack of communication between engineering contractors and the client (Lovins, 1992, p. 28):

> 'With people jumping ship, moving between contractors and so on – and the inability of contractors to retain a workforce long term, means that you'll get maybe 2 or 3 project managers and commissioning engineers working on a project...without having sufficient integration. And it's all black box stuff, so you don't know what the hell's going on...So then you end up with something where it's not really clear why it's not working. You have to go right back to first principles. And then it takes an exceptionally skilled client to do that.'[30]

Again, the primary incentives on contractors are to deliver to time and budget. They do not have incentives to optimise the performance of the building in use and their fees do not cover the cost of post-occupancy evaluations of building performance. As a result, the latter are rarely performed. A recent series of post-occupancy surveys of 'green' buildings noted the prevalence of poor functionality and user interfaces for controls, excessively complicated systems which occupants and management found difficult to use, and widespread energy inefficiency (Bordass and Bunn, 1999). As a consequence they recommended clearer communication between contractors and clients, more robust and simpler solutions, more usable controls, better support to occupiers after handover, improved monitoring and better feedback to design teams. The last is particularly important if designers are to learn from current practice and improve subsequent designs.

3. THEORETICAL INTERPRETATIONS OF THE OBSERVED BARRIERS

These observations suggest that barriers to energy efficiency in the construction process are pervasive and multifaceted. Furthermore, they seem inherent in the basic organisation of the industry and in the relationships between different actors. This puts them largely beyond the scope of traditional policy measures, such as building regulations. The latter will ensure basic standards but, in their present form, they fail to address the 'system' aspects of energy efficient design and are unlikely to encourage innovative green buildings (Gann et al., 1998).

The barriers may be interpreted in more general terms by applying the theoretical framework developed in Chapter 2. The barriers that appear of particular importance can be grouped under the headings of imperfect information, agency problems, split incentives and bounded rationality. These are discussed in turn below.

3.1 Imperfect Information

The case studies suggest that lack of information is a common obstacle to improved energy efficiency in new construction (Golove and Eto, 1996, p. 38). For example, new technology and high efficiency solutions are unfamiliar, and the presence of severe time constraints may create an incentive to stick to familiar designs. Architects lack knowledge and information on the requirements of building services. Lack of integration in the design process means that individual groups or subcontractors lack the information needed to optimise their designs. Similarly, lack of knowledge of clients' future needs creates an incentive for oversizing.

In all cases, increasing the availability, lowering the costs or improving the transfer of information would encourage more efficient designs, as would improving the capacity of actors to use information. But the feasibility of achieving this will vary. In the case of future building needs, there is an irreducible element of risk and uncertainty. In the case of new technology, there is a role for targeted information programmes, demonstration schemes and design tools to allow designers to assess the consequences of different design options. In the case of integrated design, there is scope for improving the transfer of information between different parties, but this is likely to require the replacement of contracts by alternative forms of relating and co-operating. This points to the need for organisational change within the construction process.

3.2 Agency Problems

Moral hazard can arise where the actions of one party to a contract are unobservable to the second party. Since the interests of the two parties may differ, this creates an incentive to act opportunistically after a contract has been signed.

The construction process is based on a complex series of contractual relationships, in which asymmetric information between the different parties creates numerous opportunities for moral hazard. The client is the ultimate principal, but agency relationships also exist between individual subcontractors. The case studies give a number of examples of where this acted to the detriment of energy efficiency, including the use of 'equivalent to' clauses to substitute cheaper but less efficient equipment, and the delivery of buildings with inadequate commissioning. If these problems were as easy to identify as a missing window, for example, they would be unlikely to occur. But instead, they are highly technical issues, buried deep within the building in 'black-box' services equipment and frequently missed by both the client and the Clerk of Works. The poor energy performance of building services equipment is difficult to identify since the client lacks adequate benchmarks, monitoring is inadequate and the only information that is available is the aggregate energy use of the building. If the consequences of contractor's decisions are difficult and costly to monitor and if the contractor has both a strong incentive to maximise profits and limited or zero incentive to minimise lifetime costs, opportunistic behaviour is a very likely outcome. The client's situation is not helped by the fact that post-occupancy surveys are the exception, rather than the rule.

3.3 Split Incentives

The construction process provides numerous examples of split incentive problems. We saw that: a) designers, consultants and subcontractors have no long-term interest in a building, are not liable for running costs and have an overwhelming incentive to keep within time and budget; b) there are strong incentives to cut corners to the detriment of energy efficiency in order to maximise profit margins within a highly competitive market; c) building services contractors have direct incentives to oversize equipment, due to concerns about legal liability and future flexibility; and d) fee structures provide a direct disincentive to improve energy efficiency as this means more work for less money. In all cases, one party has the information required to improve energy efficiency and both parties could gain from information sharing and the use of shared savings contracts. But the transaction costs entailed in developing such contracts may exceed the gains achievable. Against this, exhortations to consider whole-life costs are relatively weak and both the contractual organisation of the industry and the adversarial culture militate against such arrangements.

3.4 Bounded Rationality

Boundedly rational individuals require time and resources to obtain information and are limited in their ability to make full use of information. As a result, they will substitute routines and rules of thumb for more comprehensive assessments, and make sub-optimal decisions relative to the predictions of neo-classical theory. Bounded rationality may therefore create an additional barrier to energy efficiency, as well as reinforcing the operation of other barriers.

The use of rules of thumb in equipment sizing, the 'pasting' of inappropriate equipment specifications from previous contracts, the bias towards tried and trusted solutions rather than innovative energy efficient options, the dominant and negative effect of £/m^2 capital cost guidelines, the rigidity of cost plans, and the limited oversight of contractors' decisions can all be interpreted as evidence of bounded rationality. The severity of time constraints in the construction process will inevitably reduce the scope for more carefully considered decisions. Since bounded rationality is unavoidable, the relevant policy question is how to create scope for actors to make more efficient choices.

3.4 Summary

The specific problems within the construction industry can therefore be interpreted in the theoretical terms introduced in Chapter 2. A common theme is the transaction costs associated with choosing and developing energy efficient options as well as with overcoming other barriers to energy efficiency, such as split incentives. While transaction costs are incurred for all types of decision-making within the construction process, they are generally greater for energy efficient decisions than for the less efficient alternative. Hence, policy interventions that reduce transaction costs for energy efficient options may, in some circumstances, lead to public and private benefits. As an example, information market failures may begin to be overcome through requirements to assess the energy performance of buildings using standardised methodologies, or through the development of sector-specific benchmarks on building energy efficiency.

While positive transaction costs are inherent in all organisational and market arrangements, they are contingent on the nature of those arrangements and may in some circumstances be reduced by appropriate organisational change. For the construction industry, this raises the question of whether the reliance upon subcontracting through competitive tender is the most effective way to organise the construction process. Are there alternatives which reduce transaction costs and could deliver higher quality and more energy efficient buildings?

The following section highlights some relevant initiatives in the UK construction industry that may help achieve this goal.

4. THE REFORM OF THE UK CONSTRUCTION INDUSTRY

The UK construction industry is currently undergoing a number of changes as a consequence of the reform process initiated by the Latham and Egan reports. This reform process is running parallel to the development of the UK climate programme, which includes a number of initiatives to encourage energy efficiency in buildings (DETR, 2000). But to date, these processes have been almost entirely independent. Sustainability is mentioned only once in the Egan report (Egan, 1998, p. 32) and none of the seven performance targets proposed for the industry relate to building performance in use (Egan, 1998, p. 19).

But while energy efficiency and sustainability have been a side issue within this process, the reforms would appear to address many of the problems identified above. This is because they propose fundamental changes in the organisation of the industry and in the way different actors interact. There is both an opportunity and a need, therefore, to link the reform process to the wider agenda of sustainability. This requires, amongst other things: making developers more accountable for the performance of buildings in use; widespread adoption of whole-life costing; encouragement of integrated design; adoption of post-occupancy evaluation; client education; and comparison of building performance against standardised benchmarks. Three areas of reform are particularly relevant:

- partnering and integrated design;
- value management; and
- information, education and design tools.

4.1 Partnering and Integrated Design

The majority of the problems identified in the construction process derive from the contractual relationships between different firms. *Partnering* is a label used for a variety of alternative approaches to managing these relationships (Barlow et al., 1997). The emphasis in partnering is on collaboration, teamwork and learning, typically in the context of long-term relationships which are not based on formal contracts. Partnering draws its inspiration from manufacturing, where competitive tendering for suppliers is increasingly being replaced by relatively informal agreements with a few suppliers. Standards are ensured by benchmarking supplier performance and agreeing targets for continuous improvement. The aim is to achieve lower product costs, lower transaction costs and improved quality without reducing profit margins.

Partnering was a major theme of the Latham and Egan reports and is increasingly being adopted by major construction clients. There are three main forms:[31]

- *Long-term agreements between clients and contractors*: Where a major client allocates successive projects to a small number of preferred contractors on the understanding that there will be performance improvements from contract to contract.
- *Long-term agreements between main contractors and suppliers*: Where a main contractor enters into long-term agreements with subcontractors and suppliers, replacing contractual relations with a steady flow of business and a shared commitment to improving performance.

- *Project-specific partnering*: Where a main contractor attempts to introduce some of the ideas and practices into a project team that has been assembled in the traditional way for a single project.

Partnering greatly reduces the transaction costs of tendering and drawing up contracts. These are replaced by performance measurement and improvement targets for quality, timeliness and costs. Partnering also reduces the scope for opportunism, aligns the incentives of different groups and encourages information transfer.

Some of the demonstrated benefits of partnering include reduced exposure to construction and legal risk, better time and cost control, and increased opportunities for innovation through communication, trust and mutual learning (Egan, 1998; Long et al., 1999; Critchlow, 1998). When conducted properly, partnering relationships deliver better performance than those based on competitive tendering (Barlow et al., 1997). The latter has traditionally been required for public sector procurement to ensure value for money. However, current procurement guidance encourages partnering arrangements and incorporates many of the recommendations of the Egan report (HM Treasury, 2000b).

Partnering *should* offer a potential solution to many of the problems identified above. For example, it should facilitate integrated, cross-disciplinary design processes; it should reduce the incentives to cut corners to maximise profits; it should help prevent building services being 'trimmed'; and it should improve the information flows between different actors. However, evidence on the impact of partnering on energy efficiency is very limited. Analysts have focused on traditional performance measures such as construction time, not on whole-life costs (Barlow et al., 1997). As Winch notes: 'The Egan and Latham inspired agenda presently focuses entirely on the construction process rather than the constructed product' (Winch, 2000, p. 153). Also, while partnering is increasingly being adopted in the industry, it is largely confined to major construction clients. The case studies suggest that knowledge of these approaches within the higher education sector remains limited. The potential benefits of partnering will only be fully realised when the industry routinely offers the approach to smaller, occasional and less experienced clients.

4.2 Value Management and Life Cycle Costs

As with partnering, 'value management' and associated approaches have their origin in manufacturing and in the 'lean production' concepts inspired by the Toyota production system (Womack and Jones, 1996). The aim of value management is:

'... to understand a client's objectives, and meet them in the most efficient way. Value management solutions have been specifically designed to optimise the level of expenditure, whilst meeting all of the client's building requirements. This often means *simpler and more efficient installations*.'[32]

Value management involves the design team working together with a facilitator to achieve the client's needs while eliminating unnecessary costs – such as oversized equipment (Kelly and Male, 1993). The process involves identifying areas with potential for eliminating unnecessary costs; analysing the functions of that design feature; developing alternative approaches for meeting that function; evaluating their feasibility; costing selected options on a whole-life basis; and recommending an appropriate choice.[33]

Value management is thus an *integrated design process* focused on eliminating waste and using whole-life costing (Blockley and Godfrey, 2000). The last is a core requirement for green buildings, even if sustainability is not the primary driver for the adoption of value management. Partnering arrangements and value management may offer approaches that both reduce capital costs and encourage a change in attitude to whole-life costing. Furthermore, value management is easier to achieve in the context of partnering arrangements and should form an integral element of such processes.

4.3 Information, Education and Design Tools

A third theme of the Egan report is the need to improve education, training and continuous learning at all levels within the construction industry. This includes training for project managers, designers and craftsmen, together with changes in architectural education and assistance to clients in preparing briefs. A variety of initiatives have followed from this, including best-practice guidance and benchmarking initiatives from the Construction Industry Board[34] and the Building Services Industry Research Association (BSIRA).[35] However, as with partnering, sustainability and energy efficiency remain very much secondary issues. Hence, the impact of the Egan reforms is likely to remain limited until energy efficiency becomes a more prominent feature of these ongoing programmes. Specific initiatives to encourage green design could include standardised methodologies for assessing whole-life costs, design tools for analysing the energy consequences of design decisions and guidance on interdisciplinary approaches to energy efficient design.

Of particular importance are standardised and widely disseminated methodologies for assessing the energy and environmental performance of non-domestic buildings. The Building Research Establishment Environmental Assessment Methodology (BREEAM) is becoming established in the UK

(Baldwin et al., 1998), while the European Commission is proposing a co-ordinated approach, linked to standards and certification schemes (CEC, 2001). These can provide a basis for voluntary and mandatory targets, particularly for public sector buildings. There is also a need for much greater awareness of the economic benefits of energy efficient buildings among clients. This may be encouraged through demonstration projects, advice schemes and education programmes.

5. SUMMARY AND CONCLUSIONS

This chapter has demonstrated that a series of problems inhibits energy efficiency in new construction. These derive from the fragmentation of the industry, the linear design process, the reliance placed on cost-based competitive tendering and the incentives placed on different actors. The resulting problems of oversizing, reduced quality, neglect of whole-life costs and lack of integration all inhibit the development of innovative green buildings.

This chapter has also demonstrated how these problems can be interpreted in more general terms using concepts from orthodox and transaction cost economics. First, the transaction costs associated with acquiring, transferring and using information on energy efficiency are relatively high, so actors lack the information they need to optimise their designs. Second, problems of asymmetric information pervade the construction process, encouraging contractors to act opportunistically to the detriment of quality and long-term energy efficiency. Third, most parties have no incentive to minimise life-cycle costs, while the transaction costs of developing shared saving arrangements are prohibitive. And finally, in the context of severe time constraints, individuals substitute routines and rules of thumb (for example, £/m^2 guidelines) that frequently act to the detriment of energy efficiency.

While some of these problems are inherent in the nature of energy efficient buildings, a great number of them result directly from the manner in which the construction industry is structured and organised. This suggests that, while conventional energy policy measures have a role, the most promising approach to addressing the barriers is through changes in the construction industry itself. Such changes are currently underway in the UK, stimulated by the Egan report. But at present this process has practically no links with either UK climate policy or the wider sustainability agenda. This is therefore a classic illustration of the need for policy integration (Hertin and Berkhout, 2003). If the UK is to begin the transition to a low-carbon built environment, climate policy objectives must be fully integrated into this 'post Egan agenda'. At a minimum, this means increased accountability for building performance in use and the development of sustainability targets that are given equal prominence to other performance measures.

NOTES

1. An earlier version of this chapter was published in July 2003 as 'Making the link: climate policy and the reform of the UK construction industry'. *Energy Policy*, **31**(9), 865–78.
2. Assigning CO_2 emissions from electricity generation to the final user.
3. In the public and commercial sector, space heating accounts for half of delivered energy use but only one-third of CO_2 emissions (Pout et al., 1998). Lighting accounts for a quarter of CO_2 emissions and cooling for 7 per cent.
4. Energy consumption in commercial buildings increased by more than 50 per cent between 1970 and 1996, compared to a rise of only 3 per cent in public buildings (DTI, 1997). The drivers include expansion in floor space, and increased heating, lighting, IT and (particularly) air conditioning loads in individual buildings.
5. The UK government has committed to 'put the UK on a path' towards a 60 per cent reduction in CO_2 emissions by 2050 (DTI, 2003). Given the longevity of the building stock, it is natural to ask whether current building standards are consistent with this target. As an illustration, the current stock average carbon intensity of commercial offices is ~25kgC/m^2, which compares with an average of 33kgC/m^2 for new build. A pro-rata attainment of the 60 per cent target in the UK non-domestic building sector would require a stock average carbon intensity of ~10kgC/m^2 – which implies even better performance for new build.
6. The first part of this chapter uses a similar structure to Lovins (1992) and demonstrates the similarities between the US and UK construction industries.
7. Higher education case study interview, summer 1999.
8. Higher education case study interview, summer 1999.
9. Higher education case study interview, summer 1999.
10. Higher education case study interview, summer 1999.
11. Higher education case study interview, summer 1999.
12. Higher education case study interview, summer 1999.
13. Higher education case study interview, summer 1999.
14. Mechanical and electrical equipment, or building services.
15. Higher education case study interview, summer 1999.
16. Higher education case study interview, summer 1999.
17. See for example the passively ventilated structures designed by Short Architecture, London and Leicester.
18. Higher education case study interview, summer 1999.
19. Higher education case study interview, summer 1999.
20. Using 1994 data for the public and commercial sector alone, this equates to approximately 1.3MtC (Pout et al., 1998). For comparison, the UK target of 10 per cent of electricity from renewable sources by 2010 is expected to save 2.5MtC.
21. Higher education case study interview, summer 1999.
22. Higher education case study interview, summer 1999.
23. Higher education case study interview, summer 1999.
24. Higher education case study interview, summer 1999.
25. Higher education case study interview, summer 1999.
26. Higher education case study interview, summer 1999.
27. http://www.ciboard.org.uk.
28. The constraint here, as elsewhere, is the relative size of capital and energy costs. In 2000, energy costs in the university sector averaged £6.10/m^2 (HEFCE, 2001). In contrast, capital costs for new academic buildings are as high as £1500/m^2. Assuming a 30 year life, a discount rate of 5 per cent and unchanged energy prices, the estimated present

value energy costs for one case study building were ~£0.5m, compared to a capital cost of £7.2m.

29. Higher education case study interview, summer 1999.
30. Higher education case study interview, summer 1999.
31. See the Construction Best Practice Programme website: http://www.cbpp.org.uk/cbpp/.
32. http://www.bsria.co.uk/bsbpp/index.htm.
33. http://www.bsria.co.uk/bsbpp/themes/value_management/value_management.htm.
34. http://www.cbpp.org.uk/cbpp/.
35. http://www.bsria.co.uk/bsbpp/.

REFERENCES

Austin, S. and J. Steele (1999), 'Using training workshops to map interdisciplinary team working', Paper presented at Engineering in the 21st Century: The Changing World, National Conference, Harrogate, October.

Baldwin, R., A. Yates, N. Howard and S. Rao (1998), *BREEAM 98 for Offices: an Environmental Assessment for Office Buildings*, Building Research Establishment, Garston.

Barlow, J., M. Cohen, A. Jashapara and Y. Simpson (1997), *Towards Positive Partnering: Revealing the Realities in the Construction Industry*, The Policy Press, Bristol.

Blockley, D. and P. Godfrey (2000), *Doing It Differently: Systems for Rethinking Construction*, Thomas Telford, London.

Bordass, B. (2001), *Flying Blind: Everything You Wanted To Know about Energy in Commercial Buildings But Were Afraid To Ask*, Association for the Conservation of Energy, London.

Bordass, B. and R. Bunn (1999), 'The PROBE project: technical lessons from PROBE 2', Paper presented at Engineering in the 21st Century: The Changing World, National Conference, Harrogate, October.

Bowley, M. (1966), *The British Building Industry*, Cambridge University Press, Cambridge.

Brittain, J.R.J. (1997a), *Oversized Cooling and Pumping Plant*, Building Service Research and Information Association, London.

Brittain, J.R.J. (1997b), *Oversized Heating Plant*, Building Service Research and Information Association, London.

Bunn, R. and T. Smith (1999), 'Can we rethink co-ordination?', *Building Services Journal*, January, 29–32.

CEC Commission of the European Communities (2001), *Proposal for a Directive on the Energy Performance of Buildings*, COM (2001) 226 final.

Cooper, G. (1998), *Air-Conditioning America*, Johns Hopkins University Press, London.

Critchlow, J. (1998), *Making Partnering Work in the Construction Industry*, Chandos Publishing, Oxford.

DETR Department of the Environment, Transport and the Regions (2000), *Climate Change: The UK Programme*, HMSO, London.

DTI Department of Trade and Industry (1997), *Energy Consumption in the United Kingdom*, HMSO, London.

DTI Department of Trade and Industry (2003), *Our Energy Future: Creating a Low Carbon Economy*, The Stationery Office, London.

Egan, J. (1998), *Rethinking Construction*, The report of the Construction Task Force to the Deputy Prime Minister, John Prescott, on the scope for improving the quality and efficiency of UK construction, Department of Environment, Transport and the Regions, HMSO, London.

Gann, D. (2000), *Building Innovation: Complex Constructs in a Changing World*, Thomas Telford, London.

Gann, D., J. Phillimore, L. Thorburn, P. Simmonds and M. Staton (1996), 'Over-specification of engineering services', A report sponsored by the SEC Group and the DoE Construction Sponsorship Directorate, Technopolis, Brighton.

Gann, D., Y. Wang and R. Hawkins (1998), 'Do regulations encourage innovation? the case of energy efficiency in housing', *Building Research and Information*, 26(4), 280–96.

Golove, W.H. and J.H. Eto (1996), 'Market barriers to energy efficiency: a critical reappraisal of the rationale for public policies to promote energy efficiency', LBL-38059, Lawrence Berkeley Laboratory, University of California, Berkeley.

Grout, P.A. (1997), 'The economics of the Private Finance Initiative', *Oxford Review of Economic Policy*, **13**(4), 53–66.

HEFCE Higher Education Funding Council for England (2001), *Estates Management Statistics Project: Annual Report 2001*, Report 01/72, HEFCE, Bristol.

Her Majesty's Treasury (2000a), *Procurement Guide No. 2: Value for Money in Construction Procurement*, HMSO, London.

Her Majesty's Treasury (2000b), *Procurement Guide No. 4: Teamworking, Partnering and Incentives*, HMSO, London.

Her Majesty's Treasury (2000c), *Procurement Guide No. 7: Whole Life Costs*, HMSO, London.

Hertin, J. and F. Berkhout (2003), 'Analysing institutional strategies for environmental policy integration: the case of EU enterprise policy', *Journal of Environmental Policy and Planning*, **5**(1), 39–56.

Kelly, J. and S. Male (1993), *Value Management in Design and Construction: The Economic Management of Projects*, E & F.N. Spon, London.

Latham, M. (1994), *Constructing the Team*, HMSO, London.

Long, M., R. Wheddon and M. Facer (1999), 'Top Team', *Building Services Journal*, January, 22–4.

Lovins, A. (1992), 'Energy efficient buildings: institutional barriers and opportunities', Strategic Issues Paper No. 1, E Source Inc., Boulder, Colorado.

Lovins, A.B. and L.H. Lovins (1997), *Climate: Making Sense and Making Money*, Rocky Mountain Institute, Colorado.

NAO National Audit Office (1998), *The Management of Building Projects at English Higher Education Institutions*, HMSO, London.

Pout, C., S.A. Moss, P.J. Davidson, J.P. Steadman, H. Bruhns, N.D. Mortimer and J. Rix (1998), *Non-Domestic Building Energy Fact File*, BRE, Watford.

Reve, T. and R.E. Levitt (1984), 'Organisation and governance in construction', *Project Management*, 2.

Sanstad, A.H. and R.B. Howarth (1994), '"Normal" markets, market imperfections and energy efficiency', *Energy Policy*, **22**(10), 811–18.

Scrase, I. (2000), *White-collar CO_2: Energy Consumption in the Service Sector*, Association for the Conservation of Energy, London.

Wilson, A., J. Uncapher, L. McManigal, L. Hunter Lovins, M. Cureton and W. Browning (1998), *Green Development: Integrating Ecology and Real Estate*, Rocky Mountain Institute. John Wiley and Sons, New York.

Winch, G.M. (1989), 'The construction firm and the construction project: a transaction cost approach', *Construction Management and Economics*, 7, 331–45.
Winch, G.M. (2000), 'Institutional reform in British construction: partnering and private finance', *Building Research and Information*, 28(2), 141–55.
Womack, J.P. and D.T. Jones (1996), *Lean Thinking: Banishing Waste and Creating Wealth in Corporations*, Simon and Schuster, New York.

8. Barrier Busting: Overcoming Barriers to Energy Efficiency

Steve Sorrell

'Arguments that protecting the earth's climate will cost a lot rest on theoretical economic assumptions flatly contradicted by business experience... huge opportunities to save money by saving energy exist, but are being blocked by dozens of specific obstacles at the level of the firm, locality or society...Focusing private and public policy on barrier-busting can permit businesses to buy energy savings that are large enough to protect the climate and profitable enough to strengthen economic vitality, employment and competitiveness.' (Lovins and Lovins, 1997, p. i)

1. INTRODUCTION

The preceding chapters have provided detailed accounts of energy management practices and the barriers to energy efficiency within each of the case study sectors, and suggested specific initiatives at the organisational, sectoral and national level which may potentially reduce these barriers. The present chapter draws together these insights, identifies some common themes and suggests answers to the original research questions. The issues addressed are the existence and scale of energy efficiency opportunities in the case study organisations; the relative importance of different barriers to energy efficiency; the nature and operation of these barriers; and the potential for organisational, contractual and public policy measures to overcome them, including the extent to which they legitimate public intervention. The chapter also suggests some possible directions for future research.

2. WHAT IS THE SCALE OF ENERGY EFFICIENCY OPPORTUNITIES IN THE CASE STUDY ORGANISATIONS?

As the quote from Lovins and Lovins demonstrates, technology optimists claim there are 'huge' opportunities to improve energy efficiency cost-effectively. This claim is endorsed by the IPPC (2001), who define cost-effectiveness as considering capital, maintenance and energy costs alone. But this definition ignores the hidden costs associated with such opportunities, such as management and staff time.

The full extent of energy efficiency opportunities within the case study organisations could only be assessed by conducting detailed energy audits. This was not possible within the constraints of this study, and it was notable that relatively few of the case study organisations had conducted comprehensive energy audits themselves. However, the combination of questionnaire, interview and documentary evidence strongly suggests that there were a large number of 'cost-effective' opportunities (as defined above) in nearly all the organisations studied. Interviewees were nearly unanimous in agreeing that there were many opportunities with paybacks of less than 3–4 years:

> 'If I looked at all the opportunities with four year paybacks I would need another four staff…'[1]

> 'When I was working in energy management it was very easy to save my salary and somebody else's without a shadow of a doubt...'[2]

> 'I could save my wage, and [the Maintenance manager] could do the same. No problem.'[3]

> 'There are so many possible schemes you have got to start somewhere. I have a long list of potential schemes…'[4]

> 'Yes, I believe there are still lots of opportunities with three year paybacks. The problems are lack of time and lack of people…'[5]

These perceptions were confirmed by the results of energy audits at several of the case study organisations. For example, two of the UK case study universities had conducted campus-wide energy audits and each had identified a wide range of low cost energy efficiency opportunities in all the buildings examined. One university had developed plans for a £2 million investment programme over five years, all for projects with four year paybacks. Other universities were spending all of their annual energy efficiency budgets on projects with paybacks of three years or less, with many projects recovering their costs within one year. In the majority of the case study organisations the number of potential projects greatly exceeded the funds available, and was expected to continue to do so for the foreseeable future.

The postal surveys gave similar results for a wider sample of organisations. For example, 79 per cent of respondents to the UK higher education survey either agreed or strongly agreed that there were many opportunities with four year paybacks, with 14 per cent neutral and only 7 per cent disagreeing.[6] The corresponding results from the UK brewery survey were 48 per cent agreeing, 19 per cent neutral and 33 per cent disagreeing.[7] This suggests that somewhat fewer opportunities were perceived in the brewing sector than in higher education – a result that was backed up by the interviews. Since brewing is more energy-intensive, we would expect more attention to have been paid to energy efficiency.

In both brewing and higher education, interviewees commonly identified specific, low-risk measures with very high rates of return whose realisation was frustrated by a variety of obstacles. Interviewees in mechanical engineering were less specific about individual opportunities as, in most cases, they did not have the time to fully investigate them. However, they were confident that such opportunities were available and pointed to previous experience – when time constraints were less pressing – to support this judgement:

> 'One of my roles was to go round and look at the compressed air leaks, observe them, and pass it on to our maintenance department to make sure that we actually repair...The amount of leaks we had! ...But nobody is doing that now at all. Not for ten years.'[8]

Each case study investigated the extent of take-up of a number of specific energy efficiency opportunities that were recommended for the sector in best-practice literature. The results reported in Chapters 3 to 7 are comparable to those for the other countries and sectors and suggest that, in the majority of cases, a significant proportion of these opportunities had not been taken up. For example, in the UK mechanical engineering case studies, an average of 50 per cent of the suggested opportunities had not been taken up and two firms had implemented no more than a quarter. In some cases it was clear that site-specific obstacles were to blame. For example, three of the UK breweries could not use additional heat recovery as they already had more than enough hot water from the wort cooler, while another UK brewery found that the need for a new gas main made a switch to gas-fired boilers uneconomic.[9] However, in no case did these site-specific obstacles extend to more than a fraction of the suggested measures.

In sum, the case study results give strong support to the claims of Lovins and others that cost-effective energy efficiency opportunities – as defined above – are widely available. But this definition ignores hidden costs and other barriers. And as discussed below, the case study results also demonstrate that these barriers are of considerable significance.

3. WHAT ARE THE MOST IMPORTANT BARRIERS TO ENERGY EFFICIENCY IN THE CASE STUDY ORGANISATIONS?

The research employed a taxonomy of six barriers to energy efficiency: risk, imperfect information, hidden costs, access to capital, split incentives and bounded rationality. This section summarises the relative importance of each of these barriers and identifies the specific instances in which they arise, while section 4 suggests a unifying framework through which they may be interpreted.

Since the case study results were largely qualitative and required interpretation by the research teams, it was not possible to identify the relative importance of each barrier with any precision. Instead, for each sector in each country, the individual barriers were simply classified as either of high, medium or low importance. While this represents a subjective assessment by the research teams, every attempt was made to harmonise these interpretations, for example through common interview and questionnaire protocols and cross-country comparison of results. The project reports discuss this assignment in detail, while Chapters 3 to 7 highlight and discuss those barriers which were considered of particular importance in the selected country/sector cases.

Tables 8.1 to 8.3 summarise the results. Table 8.1 identifies those barriers that were considered of high importance by the research teams in each sector and country, while Table 8.2 identifies the specific instances of these barriers which were considered of high importance (for example, 'departments not accountable for energy costs' as one type of split incentive). Table 8.3 summarises the specific instances of these barriers that were reported as being of high importance by all three research teams. Together, these tables provide a useful overview of the case study results.

The results suggest that problems associated with *hidden costs* and *access to capital* appear to be the primary reason for not investing in energy efficiency in the case study organisations. Each of these is discussed below.

Table 8.1 Barriers considered to be of high importance in each sector and country

Barrier	Higher education	Brewing	Mechanical engineering	Total
Hidden costs	U G I	U G I	U G I	9
Access to capital	U G I	U G I	U G I	9
Imperfect information	U G I	U	G I	6
Split incentives	U G I	U G		5
Risk		U G	U G	4
Bounded rationality	U			1

Note: U = UK case studies; G = German case studies; I = Irish case studies.

Table 8.2 Specific instances of barriers considered to be of high importance in each sector and country

Barrier	Specific instance	HE	Brew	Mech
Hidden costs	Overhead costs of energy management	U G I	U G I	U G I
	Cost of gathering information, identifying opportunities, etc.			G I
	Cost of production disruptions	-	U	
	Loss of utility		G	
Access to capital	Capital budgeting procedures within the organisation	U G I	U I	U G I
	Availability of capital to the organisation	U G I	U G I	U
Imperfect information	Lack of information on organisational energy use	U G I	U	G I
	Lack of information on energy efficiency opportunities			G I
Split incentives	Equipment purchasers not accountable for energy costs	U I	U G	
	Departments not accountable for energy costs	U G I		
	Designers and contractors not accountable for running costs	U G I		

Barrier	Specific instance	HE	Brew	Mech
	Restricted transferability of funds	G		
Risk	Business risk		U G	U
	Technical risk associated with energy efficient technologies		U G	
Bounded rationality	Organisational routines neglect energy efficiency	U		

Note: U = UK case studies; G = German case studies; I = Irish case studies.

Table 8.3 Specific instances of barriers reported as high importance by all three research teams

Barrier	Specific instance	HE	Brew	Mech	Total
Hidden costs	Overhead costs of energy management	*	*	*	***
Access to capital	Allocation of capital within organisation	*		*	**
	Availability of capital to the organisation	*	*		**
Imperfect information	Lack of information on organisational energy use	*			*
Split incentives	Departments not accountable for energy costs	*			*
	Designers and contractors not accountable for energy costs	*			*

3.1 Hidden Costs

Hidden costs represent those costs that are not normally taken into account in engineering-economic evaluations. But to be useful, this general category needs to be disaggregated into different components, as summarised in Table 8.4.

Examples of each of these sub-categories were evident in the case studies. But in general, market and organisational transaction costs appeared to be significantly more important than either 'hidden' production costs or loss of utility. The only hidden production cost of energy efficiency that was considered important was the potential loss of production during equipment installation. But this only appeared relevant for a subset of energy efficiency opportunities in the brewing and mechanical engineering sectors and comparable problems of disruption did not appear significant in higher education. Similarly, examples of the *hidden benefits* of energy efficiency improvements outnumbered examples of the loss of utility. In the mechanical engineering sector, for example, interviewees were unable to give any examples of a loss of utility associated with energy efficient options, while interviewees in all sectors gave numerous examples of hidden benefits – such as changes to defrosting regimes at a brewery leading to energy savings, water savings, reduced maintenance and reduced deterioration of building fabric. This finding is consistent with the review of 52 manufacturing case studies by Laitmer and Finman (2000) which found that the average payback fell from four years, when only energy savings were included, to less than two years when both energy and non-energy savings were included. Similarly, there is good evidence of gains in employee productivity resulting from green building design greatly exceeding the saving in energy costs (Lovins, 1992).[10]

Table 8.4 Components of hidden costs

Sub-category	Influenced by	Examples
Hidden production costs	• Attributes of technology • Site-specific factors	Civil engineering costs; grid reinforcement costs; production interruptions; etc.
Loss of utility	• Attributes of technology • Site-specific factors	Increased noise; lower reliability; reduced service quality; etc.
Market transaction costs	• Features of primary and secondary markets for information • Organisational procedures for external transactions	Search costs for gathering and assimilating information regarding product quality; cost of specification and tendering; bargaining and negotiation costs; legal advice; etc.
Organisational transaction costs	• Organisational procedures for internal transactions	Costs of establishing, running and maintaining energy information systems; decision-making costs; monitoring and control costs, etc.

Of the different categories of market and organisational transaction costs, the factor that appeared by far the most important was the *overhead costs of energy management*, and in particular the salary costs associated with employing skilled energy management staff. In the great majority of the case studies, this was cited as the biggest obstacle to improving energy efficiency. These overhead costs proved difficult to quantify but strong evidence of their importance was provided by the severe time constraints on all respondents and interviewees in all the case study sectors. These constraints applied to all sizes of organisation, but were particularly evident for smaller mechanical engineering firms. As Box 8.1 illustrates, interviewees emphasised how time constraints were the primary reason they were unable to keep up to date with technical information, identify energy efficiency opportunities and implement energy efficiency projects.

In most cases, time constraints were considered more important than capital constraints – 'if we had more money, we wouldn't have time to spend it!'.[11] Most of the case study organisations were facing business difficulties in some form and had responded to this with drives to cut costs by reducing staff numbers. As a consequence, the already severe time constraints were widely perceived as getting worse:

> 'We used to (analyse energy data). But not now. Nobody is pushing it…We wear too many caps. You can only do the most important job they give you – until they give you a more important job.'[12]

> '...it got so that I was spending very little time on energy and it's basically eroded away. They've given me a lot of other responsibilities but they haven't given the energy management to anybody else.'[13]

> 'We run a very lean ship here...[The parent company] has benchmarks on head count...Cutting head count has been one of the main ways the company has cut costs.'[14]

In the UK, these factors had been exacerbated by declining energy prices following energy market liberalisation. This had reduced the potential cost savings from energy efficiency and consequently the priority given to energy management. Also, the available staff time was increasingly taken up with negotiating contracts for energy supply, rather than promoting energy efficiency. The liberalisation process has been proceeding more slowly in Germany and Ireland, but similar consequences could be observed there.

Box 8.1 Time constraints on energy management staff

'Why haven't we done it? It's really just time, and allocating people. We just haven't got the people...'[15]

'Our biggest problem is the lack of resources – both time and people. We spend our time firefighting...carrying out emergency repairs. This morning I had a gas leak to deal with...I wear too many hats'.[16]

'I've had some information passed on to me about the Best Practice Programme. But I just can't use it. I don't have the time If I had somebody else to actually do it, yes, I'd have no problem. But energy is a full time job, so is environmental.'[17]

'I just haven't got time to look at it...If I were to go to my mail tray now it would probably be full of documents, and the amount of time you get to go through it – well, you have to be completely ruthless. I try to pick up on the key issues of the day, and follow them through...'[18]

'I think what we are suffering from again, is the time to look at all these issues and it really comes down to manpower again. There's nobody I can delegate to. It's down to me and there is only so much work we can cover. Leaner and fitter is one word for it, but I don't think it's appropriate.'[19]

'We're very busy right now with lots of orders. It restricts the time available for studying the marketplace and looking for new innovations. We have other responsibilities...If work dropped off, then we might start looking at ways of saving energy. But then, if work dropped off there would be less money to invest...'[20]

'Our energy costs are too small for us to take much notice. We have other priorities. There is always something going on. We are always extremely busy'[21]

'I saw the information on variable speed drives in a magazine. But I don't have much time to read magazines…I am all over the brewery most of the time. There is always something going on.'[22]

The salary costs for energy management staff represent a form of hidden costs and cover activities such as operating energy information and control systems, identifying energy efficiency opportunities and specifying and tendering for capital works. Time constraints on the relevant staff are an indication that these activities are time-consuming and costly and that valuable projects may be overlooked due to lack of staff time. But since these costs are difficult to quantify, it proved hard to judge whether individual organisations were allocating sufficient staff time. Three points were clear, however. First, most organisations appeared to be allocating significantly less staff time than is recommended in best-practice publications (that is, one full-time energy manager for each £1 million of annual utilities expenditure) (Energy Efficiency Office, 1993). For example, one UK university with an annual utilities expenditure of £1.75 million, was allocating no more than 5 per cent of a single individual's time to 'core' energy management activities (Box 2.2). Second, the staff time devoted to energy management varied widely between organisations of comparable size in the same sector with comparable opportunities to improve energy efficiency. Third, the costs and benefits of staff cuts – including lost opportunities from improved energy efficiency – did not appear to have been adequately assessed. Since most of the case study organisations had not conducted comprehensive energy audits, they lacked relevant information on what these costs and benefits were. Furthermore, where energy management staff had access to this information, it was rarely communicated effectively to senior management who generally lacked interest in energy issues. As a result, the latter remained largely in ignorance of the cost-saving opportunities that were available:

> 'The main problem is a lack of understanding of savings high up in the organisation…They don't believe the message that if they put their hands in their pocket they will save in the long term.'[23]

> 'Energy efficiency is not an issue for top management. They don't want energy efficiency to do anything to disrupt research. If you want to leave all your computers on 24 hours a day, 365 days a year then that is OK. That's preferable to somebody having to wait a couple of minutes for a computer to come on.'[24]

At several points in the preceding chapters, a comparison has been made between salary overhead costs and the annual investment budget dedicated to energy efficiency projects. The argument here is that the use of stringent paybacks for such projects may potentially be justified as a means to recover salary and other overheads. It is difficult to justify loading all the salary overhead costs onto energy-saving investment projects as energy management staff are frequently performing multiple and essential tasks, which are not all expected to result in cost savings. However, the argument applies with more force to the marginal cost of increasing the staff time devoted to energy efficiency – including the cost of recruiting new staff. But the salient point is that the recovery of overhead costs was never an explicit rationale for stringent investment criteria. In higher education, for example, there were rarely any strict rules on investment criteria and interviewees made rather vague references to risk and 'good practice'. In mechanical engineering, there were more examples of explicit rules from senior management or the parent company, but in no case were overhead costs explicitly quantified and included in investment appraisal. Hence, 'hidden costs' in this instance can only influence organisational decision-making in an indirect manner through managerial decisions to reduce (or not expand) staff numbers, largely taken on the basis of inadequate assessment of the costs and benefits.

3.2 Capital Constraints

The second strong message from the research was the importance of limited access to capital. This applied at two levels: insufficient capital through internal funds combined with the reluctance to raise additional funds through borrowing or share issues; and the low priority given to energy efficiency within internal capital budgeting procedures.

The first of these applied to practically all the case study organisations and derived in part from the difficult business situation faced by many. The brewing industry was facing declining demand, overcapacity, increased competition and reduced margins and was responding with cost-cutting drives, staff reductions and site closures.[25] The mechanical engineering industry was losing export markets and was responding through short-term expediencies such as importing components from abroad, dropping product lines, laying off staff and cutting back on R&D and investment. Higher education had been underfunded for years, leading to under-investment in infrastructure, a backlog of maintenance and a funding gap for capital projects. The net result was a shortage of internal funds for capital investment in nearly all the case study organisations. This financial stringency may have contributed to the reluctance to increase aggregate borrowing, or to enter into long-term energy service contracts. However, a range of other factors contributed here, including management priorities, the perceived financial risk associated with increased gearing, the difficulty of 'ring-fencing' a revenue stream of avoided energy costs, and staff opposition to outsourcing. In addition, public sector rules led to pervasive restrictions on borrowing in the higher education sector – particularly in Germany and Ireland.

All these factors contributed to a limited pool of investment funds which was then restricted further by priority setting within internal capital budgeting procedures, coupled with rigid budgeting rules which make it difficult to transfer funds from one area to another. Since energy costs were small and energy efficiency projects could be postponed, they typically fell to the bottom of the priority list. Management attention was focused instead on strategic areas such as expansion of production, or on non-discretionary projects such as essential replacement of equipment:

> 'There is only a fixed pot of money to play with...Viable projects may not go ahead because there are preferred projects which use up the available funds. Frequently, production people come in and pinch my money...We can't see the big picture of course. Management can – they have the strategic view. And production is always paramount.'[26]

None of this implies that the case study organisations were behaving rationally in neglecting energy efficiency opportunities. On the contrary, both the wide variation in performance between the case study organisations and the existence of numerous best-practice examples in these sectors suggest that a portion of these obstacles result from potentially redeemable organisational failures. But the multiple origins of these obstacles suggest that removing them may be far from straightforward.

Capital rationing may be implemented through either strict limits on capital budgets or stringent investment criteria. While both were evident in the case study organisations, the former was generally the most important. Very few organisations came near the best-practice recommendations of a dedicated energy efficiency budget equal to 5 per cent of the annual expenditure on utilities and the majority had no such budget at all (Energy Efficiency Office, 1993). In contrast, while investment criteria were strict, they were often chosen by the engineering department itself, rather than being imposed by senior management. It was repeatedly emphasised that the majority of energy efficiency improvements resulted from new or replacement investments undertaken for reasons other than reducing energy costs, so an overemphasis on dedicated energy efficiency projects would be inappropriate. Similarly, it was generally the case that additional investment could not be undertaken without further staff resources to implement those projects. This suggests that capital constraints – whatever their origin – are largely secondary to hidden costs in inhibiting energy efficiency improvements.

3.3 Overdetermination

Hidden costs and capital constraints were considered important in nearly all the case study organisations, but they generally coexisted alongside one or more of the other barriers in our taxonomy. For example, problems of split incentives pervaded the higher education sector, both within individual universities and between those universities and external organisations such as building contractors. Within institutions, the lack of departmental accountability for energy costs reduced the incentives for individuals and departments to improve energy efficiency, while the comparable lack of accountability of building contractors contributed to oversizing of equipment, reduced quality, neglect of whole-life costs and lack of integrated design. Split incentives were particularly apparent in the German higher education sector, where they were reinforced by legal restrictions on the transfer of funds between different budgets and by complex, hierarchical and time-consuming decision-making procedures, extending as far up as the federal government.

This coexistence of multiple barriers has a cumulative effect, with the result that the neglect of energy efficiency opportunities becomes *overdetermined*. This implies that efforts to remove one barrier may only be partially successful if other barriers remain. For example, individual departments could be made accountable for energy costs through investment in sub-metering and energy information systems, combined with changes in budgeting procedures. But if the departments lack the skills, information or capital to respond to these incentives, or if time-constrained and boundedly rational individuals have more pressing priorities, the net effect may be very limited. In a similar manner, providing improved information on green building design may be of limited value if designers and contractors lack the time or skills to utilise this information, or if the client imposes strict limits on capital costs which neglect the potential for long-term revenue savings.

This demonstrates that the key issue for the barriers debate is not so much the relative importance of different barriers, but their cumulative effect. Initiatives to encourage cost-effective investments must understand and address each aspect of the problem if they are to be successful.

3.4 Differences between Countries and Sectors

Problems with hidden costs and access to capital were common to all sectors, but there were also some notable differences. For example, business risk played a substantially more important role in the private sector, while universities were constrained by rigid, public sector rules. The overhead costs of energy management were a particular constraint in mechanical engineering where SMEs could only allocate a fraction of staff time and often lacked the relevant skills. Lack of information on energy use and energy efficiency opportunities was more apparent in higher education and mechanical engineering, while the skilled engineering staff in the (more energy-intensive) brewing sector appeared much better informed. However, mechanisms for sharing information between organisations worked much better in higher education, owing to the relative absence of competitive pressures.

Split incentives were a major problem in higher education as a consequence of the large size and complex structure of the institutions, the autonomy and rivalry of individual departments, the political sensitivity of increasing costs for high profile engineering and physics departments, and the significance of equipment purchasing and operating decisions for overall energy consumption. Split incentives were less of a problem in brewing as the organisations were smaller, decisions were more centralised and cost centres were more frequently made accountable for energy consumption. In mechanical engineering, cost centres were rarely employed and the energy costs were so low that the incentive effects of departmental charging would be very limited. In general, the size and complexity of higher education institutions coupled with their multiple objectives and multiple interactions with external actors led to a greater range of barriers in this sector than in either brewing or mechanical engineering.

The case study sectors appear to face broadly similar problems in each country, so the differences between countries were less marked than those between sectors. In both brewing and mechanical engineering there was some evidence that German companies took energy efficiency more seriously (for example, longer payback periods) and performed better as a result. The reasons may lie in the political culture of the country, including greater regulatory pressure and the higher profile given to environmental issues. It was also notable that the UK mechanical engineering case studies faced greater business difficulties than those in Germany and Ireland, partly as a result of exchange rate differences. However, factors other than business risk appeared to be the greatest obstacle to improving energy efficiency in this sector.

The organisation and funding of the higher education sector differed considerably between the three countries, although the barriers to energy efficiency were similar. The German sector was characterised by centralisation of decision-making within the state government and agencies, rigid public sector rules and limited autonomy for individual universities. The net result was seriously inimical to energy and economic efficiency, but ongoing reforms to the public sector should reduce these obstacles. Funding crises were a feature of the higher education sector in all three countries and there were increasing moves towards business accounting and the use of private finance. The UK has gone furthest in this area, and initiatives such as public-private partnerships may provide a means to overcome capital constraints for major projects and encourage energy service contracting. But such partnerships are controversial and the costs and benefits are still unclear (Grout, 1997).

4. HOW BEST CAN THESE BARRIERS BE UNDERSTOOD?

The book has provided a number of complementary perspectives on the barriers debate and explored their usefulness in interpreting organisational behaviour. The taxonomy of six barriers to energy efficiency which guided the empirical research was based upon concepts which are widely used within the energy economics literature. But as was argued in Chapter 2, there is a great deal of overlap and interdependence between these categories, and each can be reinterpreted using more formal concepts from orthodox, transaction cost and behavioural economics. Table 8.5 summarises this reinterpretation.

As discussed in Chapter 2, the orthodox perspective draws a neat distinction between *market failures* within energy service markets, such as externalities and public goods, and *non-failures* where organisations are behaving rationally, given the risk-adjusted rate of return on an investment in the existing context of energy, capital and 'hidden' costs. The former may legitimate public policy intervention while the latter may not. The hidden costs are assumed to be fixed and hence not an appropriate focus of policy intervention.

This book has argued that the orthodox perspective needs to be modified in a number of ways. First the category of 'hidden' costs needs to be disaggregated into its individual elements, including the transaction costs associated with economic activity in both markets and organisations. These costs are not fixed but depend upon the particular market, organisational and contractual arrangements. Second, the importance of *organisational failures* needs to be recognised, where the barrier results from the incentives and constraints upon individual actors within an organisation and may potentially be reduced through managerial tools such as task allocation and incentive design. Third, both market and organisational failures need to be understood in relative rather than absolute terms and as a direct consequence of excessive transaction costs. Finally, the determinants of transaction costs need to be understood, including opportunism in the context of asymmetric information and the bounded rationality of individual decision-makers. Bounded rationality implies that opportunities to improve economic efficiency may be missed even when information is available and actors have the appropriate incentives.

Table 8.5 Understanding barriers to energy efficiency from orthodox, transaction cost and behavioural perspectives

Barrier	**Orthodox and agency perspectives**	**Transaction cost and behavioural perspectives**
Risk	• Energy efficient technologies may be subject to higher technical risk. • Energy efficiency investments may be illiquid and irreversible, so the value of future benefits should be discounted more highly. • Energy efficiency investments may require a high rate of return to reflect the option value of delaying investment.	• Energy efficiency investments may have a high asset specificity and hence carry higher risks than other forms of investment. • Individual and organisational perceptions of risk may depart from those assumed in orthodox models and the resulting loss aversion may create a bias against energy efficient investment.
Imperfect information	• Some types of information relevant to energy efficiency may have the features of a public good and hence may be under-supplied by markets. • Since energy efficiency has the characteristics of a credence good, while energy commodities have the characteristics of a search good, there may be a systematic bias against the former. • The search costs associated with identifying the energy consumption of products may be high, thereby creating a barrier to such purchases. • In some circumstances, asymmetric information within energy service markets may lead to the adverse selection of inefficient products.	• There will be transaction costs associated with acquiring, understanding and applying information. Bounded rationality ensures that these will be significant, even when information is freely available. • The transaction costs of information acquisition may be high as a consequence of poor presentation, the lack of credibility of the source or the absence of interpersonal contacts.

Hidden costs	• Some energy efficient options may be associated with hidden costs, such as disruptions to production. • Some energy efficient technologies may perform poorly along other dimensions, such as reliability. • The search costs associated with identifying energy efficient products may be high. • Energy efficiency investments may need to recover 'overhead' costs.	• The transaction costs associated with maintaining information systems, conducting energy audits, identifying opportunities, tendering, selecting suppliers, seeking approval for capital expenditures and so on may outweigh the potential savings in energy costs.
Access to capital	• The cost of obtaining additional capital may exceed the average cost of the existing debt/equity mix, owing to the financing risk of increased gearing. • Agency problems between shareholders and managers, plus the signalling associated with external financing may create a preference for internal finance and a reluctance to increase borrowing. • Agency problems between central and divisional managers may lead to capital rationing as a form of control. • There may be asymmetry between the individual risks and rewards of energy efficiency projects.	• There will be transaction costs associated with obtaining additional funding from either internal or external sources, and these may be more significant for small, cost-saving projects. • The transaction costs of transmitting and assessing information on investment opportunities may inhibit optimal decision-making. • Given severe constraints on time and attention, managers may focus on strategic investments and overlook small cost-saving opportunities
Split incentives	• For a range of reasons, individuals, departments or organisations may not be able to appropriate the benefits of energy efficiency investments.	• Transaction costs inhibit the development of shared savings contracts to overcome split incentive problems.

Barrier	Orthodox and agency perspectives	Transaction cost and behavioural perspectives
Bounded rationality	N/A	• Constraints upon time, attention, resources and ability to process information lead to the use of imprecise routines and rules of thumb, which may systematically neglect the small cost savings from energy efficiency improvements. • Loss aversion and status quo bias contribute to inertia and the undervaluing of the benefits of energy cost saving relative to out-of-pocket costs of investment. • Risk aversion with respect to gains reinforces this inertia.

Transaction cost economics (TCE) can provide a unifying framework which encompasses the majority of the barriers in Table 8.5 and subsumes both the orthodox market failures and the different types of organisational failure. For example, split incentives persist because transaction costs inhibit the development of shared savings contracts; the transaction costs associated with acquiring, understanding and applying information lead consumers to make sub-optimal decisions based on imperfect information; the transaction costs of monitoring and control within internal agency relationships lead to a second-best imposition of strict investment criteria; the organisational transaction costs associated with establishing and maintaining energy information systems justify the use of strict investment criteria to recover those costs; and so on.

Since individuals are boundedly rational, transaction costs are unavoidable. But the level of these costs is contingent upon the particular form of economic organisation, and since there is no reason to suppose that current organisational arrangements are optimal it is possible that alternative institutional arrangements could reduce transaction costs and increase social welfare. The distinction between market/organisational failures and non-failures is then transformed into a distinction between *avoidable* and *unavoidable* transaction costs. Arguments over the size of the energy efficiency gap then reflect different views on which of these costs can be avoided.

Since transaction costs are notoriously difficult to measure, TCE focuses upon their determinants (for example, complexity, frequency asset specificity) and explores how this influences the choice between different types of governance structure, such as markets and hierarchies. This approach could be applied to the barriers debate. For example, differences in the asset specificity, frequency of purchase and complexity of different energy efficient technologies (for example, condensing boilers versus CHP) could be used to assess the relative appropriateness of in-house provision versus external contracting (Ostertag, 2003). If the latter scores well, it may provide an attractive, private sector route to overcoming the barriers to energy efficiency for certain types of technology.

However, this macro-level choice between different types of governance structure tells us little about micro-level issues such as the relative merits of different types of organisational incentives and routines. While agency theory may contribute here, it tends to be too abstract and mathematical to be of practical value for either business strategy or public policy. Instead, more relevant insights may be obtained from the business management literature (for example, Florida, 1996), best-practice publications (for example, Energy Efficiency Office, 1993) and empirical studies of successful organisations (for example, Romm, 1999; Porter and van der Linde, 1995). This book has tried to combine these insights with more formal concepts from economic theory, but TCE was never developed with the needs of managers and practitioners in mind. Hence, there is much work to be done on integrating these approaches and it is likely that TCE will need to be complemented with other ideas before it can be of greater value.

A particular weakness of TCE is its limited treatment of bounded rationality. As argued in Chapter 2, TCE incorporates bounded rationality as a background assumption, but does not explore its full implications for individual decision-making (Foss, 2003). It was for this reason that behavioural economics was incorporated into the theoretical framework of Chapter 2 and used to highlight the importance of individual and organisational routines, loss aversion and inertia. Following de Almeida (1998), the empirical research employed 'bounded rationality' as a separate category of barrier, but in practice this proved difficult to investigate. Limitations upon individual decision-making were present in all circumstances and merely served to compound the neglect of energy efficiency that was encouraged by other factors such as time constraints and lack of information. Similarly, while inertia and the 'status quo bias' may have been a feature of individual decision-making, it was rarely possible to explore this properly within the constraints of interview-based case study research. This suggests that further work in this area will need to involve more micro-level studies of the determinants of individual decisions and/or the evolution of individual and organisational routines. Such work could potentially draw upon the long-standing but currently neglected tradition of 'behavioural research' in energy efficiency, which incorporates insights from social psychology (Stern, 1992; Lutzenheiser, 1993; Stern and Aronsen, 1984).[27]

5. HOW CAN THESE BARRIERS BE OVERCOME?

5.1 The Rationale for Policy Intervention

Barriers can be overcome if there exist alternative institutional arrangements that can lower the net transaction cost of improving energy efficiency. This may be achieved through initiatives by the end-user, such as recruiting an energy manager; initiatives by actors within the energy service market, such as the expansion of energy consultancy activities; or initiatives by government, such as a labelling scheme for electric motors. TCE advocates a comparative approach in which the relative merits of different types of governance structure for different types of transaction are assessed. The most appropriate form will vary with context and with different types of end-use technology. While TCE asserts that competitive pressures will lead to an efficient choice of governance structures, it does not assume that all existing structures are optimal or that there is no room for improvement.

The polarised views of 'technological optimists' and 'economic pessimists' on the size of the efficiency gap are complemented by equally polarised views on the appropriate role of public policy – from laissez-faire market liberalism at one extreme to highly interventionist approaches on the other. But since market, government and organisational co-ordination are each associated with transaction costs, each is equally prone to failure. Hence, while there is undoubtedly a role for public policy, each proposal needs to be assessed on its merits, based on an estimate of the costs and benefits. Unfortunately, difficulties in making these estimates, combined with the relative absence of credible ex-post evaluations of previous energy efficiency programmes, creates the space for political debates fuelled largely by ideology and special interests.

Following the TCE logic, any organisational or public policy initiative to improve energy efficiency cost-effectively should aim to:

- *overcome information market failure*: provide accurate, trustworthy and easily assimilated information on energy efficiency to redress the disparity with information on energy supply;
- *economise on bounded rationality*: allow actors to make energy efficient choices without requiring extensive effort in gathering and analysing information;
- *align incentives*: ensure, as far as possible, that the incentives of different groups are complementary and act in the direction of improved energy efficiency; and
- *safeguard against opportunism*: ensure, as far as possible, that asymmetric information does not encourage decisions or actions that undermine energy efficiency.

Public policies that meet these criteria may be targeted at market or organisational failures at any stage of the supply chain for energy services, including the end-user. While organisational failures are primarily the responsibility of the individual organisation, TCE suggests that there may be circumstances where public policy can assist organisations in overcoming these problems. Two very good examples of this are provided by Howarth et al. (2000) in their review of the US Green Lights and Energy Star Office programmes.

In the Green Lights programme, organisations in the public and private sector enter into voluntary agreements with the US Environmental Protection Authority (USEPA) which commit them to implementing energy efficient lighting upgrades that satisfy conventional investment criteria while maintaining or improving lighting quality. In return, the USEPA provides information, software and technical assistance and oversees compliance. This programme has been a great success, saving $440 million in energy costs over a six-year period at an average annual rate of return of 45 per cent (USEPA, 1997). Howarth et al. attribute the programme's success to its ability to overcome organisational failures, such as asymmetric information between senior managers and individual departments, and split incentives between production and facilities staff:

> 'The Green Lights programme...aims to introduce credible data on relevant investment opportunities at a crucial point in a firm's organisational chart...In effect, EPA certifies the economic value of products and services with desirable environmental characteristics, short circuiting the principal-agent problem within firms by providing effective high quality information at the level of the firm that is most relevant to decision-making...'

> 'The same policy mechanism that addresses the agency problem also addresses the split-incentives problem. The decision-making level within the firm's hierarchy that maximises firm-wide profits for energy efficiency investments is where facilities management and production budgets are combined on the same ledger. This may be as high as the office of the chief financial officer. Without outside intervention, this level of the firm typically would not evaluate lighting retrofits or similar decisions. Such decisions would be left to appropriate divisions, where a negative decision likely would be reached. By focusing attention and providing credible data on products and services. Green Lights contributes to the resolution of this organisational failure.' (Howarth et al., 2000, p. 483)

In the Energy Star Office programme, the USEPA oversees the Energy Star labelling scheme, publicises the efforts of industry groups and encourages consumer demand through extensive advertising. This programme has also been a great success (USEPA, 1998) and has created a de facto international standard for the energy performance of computers and other office products. From the perspective of the equipment purchaser, the programme overcomes information barriers and economises on bounded rationality by providing assurance that the purchased equipment is energy efficient. This avoids the need to develop detailed and time-consuming procurement procedures to select energy efficient models. However, the aim and impact of the programme is wider than this:

'Viewed superficially, the programme might be interpreted merely as a means of signalling to consumers that a product meets EPA's standards of cost effectiveness and environmental performance. From this perspective, EPA's role is to provide neutral and hence credible information that cuts through informational asymmetries with respect to the product quality claims of self-interested sellers. Viewed more broadly, however, the aim of the Energy Star programme is to transform the entire cycle of product design, manufacturing and marketing so that energy efficiency – previously a negligible factor in the market for office equipment – becomes a central concern of corporate decision-makers. As with Green Lights, Energy Star targets crucial levels of the organisational chart to effect pervasive changes in prevailing behavioural patterns.' (Howarth et al., 2000, p. 484)

Howarth et al. (2000, p. 485) argue that the net benefits of both programmes are likely to be large as the administrative costs are relatively small. However, this judgement is based upon anecdotal evidence and there has yet to be a comprehensive assessment of the public and private costs and benefits. This is a common feature of energy efficiency policy and hence an issue of concern. One of the few areas where the costs and benefits of energy efficiency policies have been closely examined is US utility demand-side management (DSM) programmes, which Joskow and Marron (1993) argued were not cost-effective. This was subsequently challenged by Levine and Sonnenblick (1994) among others, who pointed out that the programmes had multiple objectives as they were targeted at low income households. Other relevant studies include Levine et al. (1994), who found that US appliance standards had a total benefit-to-cost ratio of more than 2.5 (at a discount rate of 6 per cent). Generally, however, this type of study is difficult to conduct and examples are rare, with the result that views on the costs and benefits of different policy interventions lack a firm empirical basis.

The need for a policy mix

A clear conclusion from the empirical research is that barriers to energy efficiency are multi-faceted, diverse and often specific to individual technologies and sectors. This implies that effective policy solutions will need to address the particular features of individual energy service markets, the circumstances of energy-using organisations, and the multiple barriers to energy efficiency within each. As a result, it is likely that a *policy mix* will be required, in which several different initiatives can work together in synergy. For example, while carbon taxes may create price incentives to improve energy efficiency, the response will be muted in many sectors unless steps are taken to lower transaction costs. Conversely, if such steps are not taken, carbon pricing may need to be unacceptably high to have a significant impact on energy demand.

This argument for a co-ordinated policy mix is increasingly supported by contemporary research on environmental policy, including the comprehensive study by Gunningham and Gabrosky (1998). In the case of energy efficiency, the policy mix is likely to include:

- *education and information programmes*: including best-practice schemes, demonstration projects and training initiatives;
- *voluntary approaches*: including market transformation initiatives and voluntary agreements with end-users;
- *economic instruments*: including carbon taxes, emissions trading, and tax allowances for investment in energy efficient technologies; and
- *traditional regulations*: including public sector accounting rules, building regulations and minimum standards for the energy efficiency of equipment.

For each of the country/sector studies in this project we have indicated a range of possible policy initiatives at the sector and national level. This gives a wide range of suggestions and ideas that are based on the current circumstances in each sector and country. These suggestions are indicative, since it was beyond the scope of the project to identify an 'optimal' mix for each sector, or to assess the likely efficiency and effectiveness of individual measures. The results do, however, illustrate the diversity of approaches that are likely to be required. By implication they also illustrate the importance of co-ordination in ensuring that individual policies complement one another rather than conflict (Sorrell, 2003; Sorrell et al., 2003). Table 8.6 illustrates this by summarising the suggestions for one of the project sectors – higher education in the UK.

Table 8.6 A policy mix to encourage energy efficiency in the UK higher education sector

Category	Examples
Education and information	• Best-practice information on generic and sectoral energy efficiency opportunities. • Best-practice information on the benefits of energy service contracting. • Best-practice information on value management and integrated design. • Best-practice information on partnering initiatives in

Category	Examples
	construction procurement. • Best-practice information on whole-life costing in public sector purchasing procedures. • Development of model contracts for energy service contracting. • Promotion of information sharing initiatives such as the Energy Share Fair Network. • Development of standardised methodologies for estimating the energy and environmental performance of buildings (e.g. BREEAM). • Development of standardised benchmarks for the energy performance of university buildings. • Labelling schemes for key items of equipment such as computers and photocopiers. • Education and training initiatives for university staff and construction industry professionals. • Demonstration projects and design advice schemes.
Voluntary approaches	• Environmental management schemes such as EMAS. • Making a Corporate Commitment Campaign. • Voluntary adoption of GHG targets by universities, administered through HEFCE. • Voluntary agreements on energy efficiency with equipment manufacturers.
Economic instruments	• Carbon/energy taxes such as the Climate Change Levy. • Carbon emissions trading schemes, such as the EU ETS. • Capital allowances for investment in qualifying energy efficient equipment.
Traditional regulations	• Relaxation of the HEFCE rules on university borrowing. • Inclusion of energy management practices and whole-life costing within HEFCE audit procedures. • Introduction of reporting requirements on building energy consumption and comparison against benchmarks. • Simplification of public procurement rules to facilitate energy service contracting. • Strengthening of employment protection rules to facilitate energy service contracting. • Tightening of energy efficiency standards within building regulations (new and refurbishment). • Minimum standards for the energy efficiency of equipment (e.g. motors).

6. DIRECTIONS FOR FURTHER RESEARCH

While there seems little prospect of resolving the barriers debate to everyone's satisfaction in the near future, the increasing political importance of reducing CO_2 emissions should encourage greater attention to the problem and lead to a more sophisticated understanding of the issues involved. As this book has argued, it is time to transcend the dichotomy of 'engineering optimist' versus 'economic pessimist' and instead develop an understanding that is informed both by contemporary developments in economic theory and detailed empirical observations of particular energy service markets. Such research should be able to identify where public policies to encourage energy efficiency are likely to be appropriate, and to inform their design.

There are a number of areas where further research would be valuable. First, at the organisational level, it would be useful to complement the present study by focusing on the determinants of success in energy management, rather than the reasons for failure. This follows the long tradition of best-practice literature and could draw upon the expanding work in environmental management (Welford and Starkey, 2001). Such research would benefit from more objective measures of 'success' in energy management, including comparison with sector-specific benchmarks. The relative paucity of data in this area is a hindrance to practical energy management and needs to be addressed.

Second, there is a need for more research on the *supply chain* for different energy services, including the relationships between the different actors involved. This recognises the limitations of focusing solely on the end-user and the importance of market transformation as a policy response. Chapter 7 of this book provided an example of this approach for non-domestic buildings and indicated the complexity of the issues involved. There is scope for more research in this area (for example, in identifying the impact of partnering arrangements on energy efficiency) as well as on specific technologies such as office equipment and lighting.

Third, there is need for more work on the determinants of transaction costs in different energy service markets and the relative merits of different institutional arrangements. An obvious focus for such an approach would be the potential for energy service contracting: for example, which types of sectors, technologies and sizes of organisation is this appropriate for? Such work could be informed by the growing body of literature on outsourcing IT support and other non-core functions (for example, Hancox and Hackney, 1999) and would benefit from a quantitative, survey-based approach that developed more precise measures for the determinants of transaction costs in different circumstances – such as uncertainty and asset specificity.

Fourth, there is need for more work on the costs and benefits of previous energy efficiency programmes and the likely efficacy of proposed policy options. Both are challenging methodologically, but there is a growing body of experience in OECD countries from which to draw.

Finally, there is a need for fuller appraisal of the interactions between different instruments – both within climate policy and between climate policy and other policy areas. These interactions are likely to have a determinate impact on the success of policy initiatives, but at present the topic of policy interaction remains seriously under-researched (Sorrell et al., 2003). One objective for this research could be to provide useful guidance on the effective design of policy packages targeted at particular energy service markets.

Research in all these areas could benefit from a pluralist approach to the selection and use of theoretical ideas, both from within economics and from other social science disciplines. This is because while each approach has its merits it also has corresponding weaknesses. For example, while TCE provides valuable insights into the choice between different governance structures, it is of much less value in understanding organisational incentives, culture and routines and hence in drawing conclusions relevant to organisational policy. Similarly, it offers few insights into the innovation and diffusion of new technologies and pays insufficient attention to organisational learning. These weaknesses could potentially be overcome by incorporating ideas from other areas, such as the 'organisational capabilities' approach pioneered by David Teece (Teece and Pisano, 1994) and the evolutionary economics pioneered by Nelson and Winter (1982). In drawing upon these different approaches, the goal should not be to develop an all-encompassing, unified theory, but to develop complementary insights which may be more or less applicable in different circumstances.

There is clearly more work to be done in understanding barriers to energy efficiency and in identifying their determinants in different contexts. But while this is an appropriate objective for academic research, the ultimate objective for climate policy is to overcome these barriers. Research should aim to contribute to both objectives. If the threat of climate change is real, this is an urgent and critically important task.

NOTES

1. UK university case study interview, summer 1999.
2. UK mechanical engineering case study interview, summer 2000.
3. UK mechanical engineering case study, summer 2000.
4. UK university case study interview, summer 1999.
5. UK brewery case study interview, summer 1999.

6. Eighty-three per cent of the respondents had conducted energy audits to identify these opportunities.
7. Fifty-six per cent of the respondents had conducted energy audits to identify these opportunities, including 85 per cent of the large sites.
8. UK mechanical engineering case study interview, summer 2000.
9. These difficulties were more apparent in the brewing sector because: first, energy use in breweries is more process-specific and opportunities are less uniform than in the other sectors; and second, many of the more common energy efficiency measures have already been applied in breweries so those technologies which have not been adopted are more variable in their suitability.
10. Case studies suggest that the better visual, acoustic and thermal comfort in well-designed, energy efficient buildings can improve labour productivity by as much as 16 per cent (Lovins, 1994). Since labour costs in commercial buildings are typically 25 times greater than energy costs, these benefits can dwarf those from energy savings.
11. UK university case study interview, summer 1999.
12. UK mechanical engineering case study interview, summer 2000.
13. UK mechanical engineering case study interview, summer 2000.
14. UK mechanical engineering case study interview, summer 2000.
15. UK mechanical engineering case study interview, summer 2000.
16. UK mechanical engineering case study interview, summer 2000.
17. UK mechanical engineering case study interview, summer 2000.
18. UK mechanical engineering case study interview, summer 2000.
19. UK mechanical engineering case study interview, summer 2000.
20. UK brewery case study interview, summer 1999.
21. UK brewery case study interview, summer 1999.
22. UK brewery case study interview, autumn 1999.
23. UK university case study interview, summer 1999.
24. UK university case study interview, summer 1999.
25. The severity of these difficulties varied between the three countries, being worst in the UK. The same comments apply to mechanical engineering.
26. UK mechanical engineering case study interview, summer 2000.
27. This work is dominated by studies of household decision-making. But as Lutzenheiser notes (1993, p. 276): 'The available research, and the literature on organisational behaviour, suggests that a simple rational model of energy use and conservation decision making applies no better to organisations than it does in the residential sector. A significant amount of useful research would also seem possible in this area'.

REFERENCES

de Almeida, E.L.F. (1998), 'Energy efficiency and the limits of market forces: the example of the electric motor market in France', *Energy Policy*, **26**(8), 643–53.

Energy Efficiency Office (1993), 'Organisational aspects of energy management', General Information Report 12, ETSU, Harwell.

Florida, R. (1996), 'Lean and green: the move to environmentally conscious manufacturing', *California Management Review*, **39**(1), 80–175.

Foss, N.J. (2003), 'Bounded rationality in the economics of organisation: "Much cited and little used"', *Journal of Economic Psychology*, **24**, 245–64.

Grout, P.A. (1997), 'The economics of the private finance initiative', *Oxford Review of Economic Policy*, **13**(4), 53–66.

Gunningham, N. and P. Gabrosky (1998), *Smart Regulation: Designing Environmental Policy*, Clarendon Press, Oxford, UK.

Hancox, M. and R. Hackney (1999), 'Information technology outsourcing: conceptualising practice in the public and private sector', Proceedings of the 32nd Hawaii International Conference on System Sciences, Hawaii.

Howarth, R.B., B.M. Haddad and B. Paton (2000), 'The economics of energy efficiency: insights from voluntary participation programmes', *Energy Policy*, **28**(6-7), 477–86.

IPCC (Intergovernmental Panel on Climate Change) (2001) *Climate Change 2001: Mitigation*, Cambridge University Press, Cambridge.

Joskow, P. and D. Marron (1993), 'What does a negawatt really cost: evidence from utility conservation programmes', *The Energy Journal*, **13**(4), 31–74.

Laitmer, J.A. and H. Finman (2000), 'Productivity benefits from industrial energy efficiency investments', a monograph prepared for the EPA Office of Atmospheric Programmes, Washington, DC.

Levine, M. and R. Sonnenblick (1994), 'On the assessment of utility demand-side management programmes', *Energy Policy*, **22**(10), 848–856.

Levine, M.D., E. Hirst, J. Koomey, J. McMahon and A. Sanstad (1994), *Energy Efficiency, Market Failures and Government Policy*, Lawrence Berkeley Laboratory and Oak Ridge Laboratory, Berkeley, CA.

Lovins, A. (1992), 'Energy efficient buildings: institutional barriers and opportunities', Strategic Issues Paper No. 1, E Source Inc., Boulder, Colorado.

Lovins, A.B. and L.H. Lovins (1997), *Climate: Making Sense and Making Money*, Rocky Mountain Institute, Old Snowmass, Colorado.

Lutzenheiser, L. (1993), 'Social and behavioural aspects of energy use', *Annual Review of Energy and the Environment*, **18**, 247–89.

Nelson, R. and S. Winter (1982), *An Evolutionary Theory of Economic Change*, Belknap Press, Cambridge.

Ostertag, K. (2003), *No Regrets Potentials in Energy Conservation: An Analysis of the Relevance, Size and Determinants*, Physica-Verlag, Heidelberg, New York.

Porter, M.E. and C. van der Linde (1995), 'Toward a new conception of the environment-competitiveness relationship', *Journal of Economic Perspectives*, **9**(4), 97–118.

Romm, J.J. (1999), *Cool Companies: How the Best Companies Boost Profits and Productivity by Cutting Greenhouse Gas Emissions*, Island Press, Washington DC.

Sorrell, S. (2003), 'Carbon trading in the policy mix', *Oxford Review of Economic Policy*, **19**(3), pp. 420–37.

Sorrell, S., C. Boemare, R. Betz, D. Haralampopoulos, P. Konidari, D. Mavrakis, C. Pilinis, P. Quirion, J. Sijm, A. Smith, S. Vassos and R.Walz (2003), 'Interaction in EU climate policy', final report to DG Research under the Framework V project 'Interaction in EU Climate Policy', SPRU (Science and Technology Policy Research), University of Sussex, Brighton.

Stern, P.C. (1992), 'What psychology knows about energy conservation', *American Psychologist*, **47**, 1224–32.

Stern, P.C. and E. Aronsen (eds) (1984), *Energy Use: The Human Dimension*, W.H. Freeman, New York.

Teece, D. and G. Pisano (1994), 'The dynamic capabilities of firms: an introduction', *Industrial and Corporate Change*, **3**(3), 537–56.

USEPA (United States Environmental Protection Authority) (1997), *Building on Our Success: Green Lights and Energy Star Buildings 1996 Year in Review,* Office of Air and Radiation, Washington.

USEPA (United States Environmental Protection Authority) (1998), Energy Star Office Product Program, Atmospheric Pollution Prevention Commission, Washington DC.

Welford, R. and R. Starkey (eds) (2001), *The Earthscan Reader in Business and Sustainable Development*, Earthscan, London.

Annex A: Energy Management Questionnaire for the Mechanical Engineering Sector

Thank you for participating in this international research project funded by the European Commission (DG-XII). Within this project, research teams from the UK, Ireland and Germany are exploring why organisations do or do not make cost-effective energy efficiency investments. This project looks at both public and private sectors and seeks to identify how government policy can encourage energy efficiency.

Prior to my visit, I would like to ask a few general questions on your company, its current level of energy consumption, and the energy efficiency measures that have been realised. **The questionnaire should take no more than 15-20 minutes to complete**. If the information is not known or not readily available then please leave the section blank.

I guarantee the confidentiality of all data and information. The draft results of this research will be sent to all participating organisations and full opportunity will be given for comment. The anonymity of the participating organisations will be preserved in all resulting reports and publications.

Thank you very much for taking the time to answer this questionnaire. Please send the completed forms back in the enclosed envelope.

1. Identification

Company..

Name...

Position..

2. Key individuals

Could you please indicate those individuals in your company who have an influence on energy decision-making and who may be potential candidates for an interview

Name	Position

3. The company

Approximate number of employees ....

Approximate annual turnover (i.e. sales) ....

(leave blank if information not known)

4. Annual energy use

Please indicate your company's approximate annual consumption of:

Fuel.. Electricity....................................

(please indicate units)

Please indicate your company's approximate annual expenditure on:

Fuel.. Electricity.........................

(leave blank if information not known)

5. Contract energy management

Do you use contract energy management? Yes ☐ Nc ☐
If yes, could you briefly indicate the coverage of the contract:

....

....

....

6. Energy information systems

At what level is energy use generally metered?

	Site?	Building?	Individual equipment?
Electricity	☐	☐	☐
Steam/hot water	☐	☐	☐

How frequently is energy use generally recorded?

	Annually?	Monthly?	Weekly?	Daily?
Electricity	☐	☐	☐	☐
Fuel	☐	☐	☐	☐

Do you monitor trends in energy consumption? Yes ☐ No ☐
Are consumption records adjusted for weather conditions? Yes ☐ No ☐
Is a monitoring and targeting scheme employed? Yes ☐ Nc ☐
Is energy performance advised to staff? Yes ☐ No ☐
Are cost centres charged for the energy they consume? Yes ☐ No ☐
Is consumption compared with benchmarks? Yes ☐ No ☐
Have you conducted energy audits? Yes ☐ No ☐

7. Energy efficiency opportunities

How much do you agree with the following statement:

'There are a wide range of energy efficiency measures that could be implemented in my company that would yield paybacks of less than three years at current energy prices'

Strongly agree	Agree	Neutral	Disagree	Strongly disagree	Don't know
☐	☐	☐	☐	☐	☐

8. Information sources on energy efficiency opportunities

Please rank the usefulness of each of the following information sources on energy efficiency opportunities:

Source	**Excellent**	**Good**	**Average**	**Poor**	**Don't use**
Colleagues within the company	☐	☐	☐	☐	☐
Network of contacts in the sector	☐	☐	☐	☐	☐
Energy manager groups/networks	☐	☐	☐	☐	☐
DETR Best Practice Programme	☐	☐	☐	☐	☐
Professional associations	☐	☐	☐	☐	☐
Trade/Technical journals	☐	☐	☐	☐	☐
Technical conferences/seminars	☐	☐	☐	☐	☐
Energy supply industry	☐	☐	☐	☐	☐
Equipment suppliers	☐	☐	☐	☐	☐
Consultants	☐	☐	☐	☐	☐

9. SELF-ASSESSMENT PROFILE OF ENERGY MANAGEMENT

This page is a 'self-assessment' exercise developed by the UK Energy & Environmental Management Division. **Reading down each column** in turn, could you please circle the box that corresponds most closely to current practice in your company.

	Policy	Organising	Information systems	Awareness	Investment
0	No explicit policy	No energy management or any formal delegation of responsibility for energy consumption	No information system. No accounting for energy consumption	No promotion of energy efficiency	No investment in energy efficiency
1	Unwritten set of guidelines	Energy management the part-time responsibility of someone with only limited authority or influence	Cost reporting based on invoice data. Engineer compiles reports for internal use within technical department	Informal contacts used to promote energy efficiency	Only low cost measures taken

2	Unadopted energy policy set by energy manager or senior departmental manager	Energy manager reports to ad-hoc committee, but line management and authority are unclear	Monitoring & targeting reports based on supply meter data. Energy unit has ad-hoc involvement in budget setting	Some ad-hoc staff awareness training	Investment using short term payback criteria only
3	Formal energy policy, but no active commitment from top management	Energy manager accountable to energy committee representing all energy users in the organisation	M&T reports for individual premises based on sub-metering, but savings not reported effectively to users	Programme of staff awareness and regular publicity campaigns	Same payback criteria as for all other investment
4	Energy policy, action plan and regular review with commitment of top management.	Energy management fully integrated into management structure. Clear delegation of responsibility for energy consumption	Comprehensive system sets targets, monitors consumption, identifies faults, quantifies savings and provides budget tracking	Marketing the value of energy efficiency and the performance of energy management both within and outside the organisation	Positive discrimination in favour of 'green' schemes with detailed investment appraisal of all newbuild and refurbishment opportunities

10. Implementation of energy efficiency measures

The following tables list some common measures for reducing energy consumption. **Please indicate the extent to which your company has implemented each measure by assigning it a number on a scale from 1 (not implemented) to 5 (extensively implemented).**

	1	2	3	4	5
Building fabric					
Draughtproofing windows, doors & rooflights?	☐	☐	☐	☐	☐
Fitting windows with double or secondary glazing?	☐	☐	☐	☐	☐
Fitting door closers to external doors?	☐	☐	☐	☐	☐
Use of plastic or forced air curtains in loading bays?	☐	☐	☐	☐	☐
Installation of fans in high ceilinged rooms to reduce temperature gradient?	☐	☐	☐	☐	☐
Retrofitting insulation to walls and roofs?	☐	☐	☐	☐	☐
Boiler plant & space heating					
Insulation of distribution pipes, valves and flanges?	☐	☐	☐	☐	☐
Use of boiler sequencing controls?	☐	☐	☐	☐	☐
Replacement of central generation of hot water with point of use application?	☐	☐	☐	☐	☐
Installation of thermostatic radiator valves?	☐	☐	☐	☐	☐
Programming HVAC controls to match occupancy patterns?	☐	☐	☐	☐	☐
Use of weather compensation & optimum start controls?	☐	☐	☐	☐	☐
Use of Building Energy Management Systems (BEMS)?	☐	☐	☐	☐	☐

Lighting					
Replacement of 38mm fluorescents with 26mm?	☐	☐	☐	☐	☐
Use of high frequency fluorescents in new & replacement fittings?	☐	☐	☐	☐	☐
Replacement of tungsten filament lamps with compact fluorescents?	☐	☐	☐	☐	☐
Replacement of fluorescents with SOX/SON discharge lighting?	☐	☐	☐	☐	☐
Use of localised task lighting in preference to general lighting?	☐	☐	☐	☐	☐
Installation of time controls with manual override?	☐	☐	☐	☐	☐
Use of photocell, acoustic or movement sensors?	☐	☐	☐	☐	☐
Integration of lighting controls into BEMS?	☐	☐	☐	☐	☐
Compressed air					
Switch off when there is no demand for air?	☐	☐	☐	☐	☐
Regular inspection & elimination of leaks?	☐	☐	☐	☐	☐
Regular maintenance & condition monitoring?	☐	☐	☐	☐	☐
Generation of compressed air at the minimum required pressure?	☐	☐	☐	☐	☐
Consideration of energy consumption & part load operation when purchasing new compressors?	☐	☐	☐	☐	☐
Pre-cool inlet air or duct air from outside?	☐	☐	☐	☐	☐
Compressor heat recovery?	☐	☐	☐	☐	☐

Heat treatment

Insulation of furnaces to economic thickness?	☐	☐	☐	☐	☐
Accurate control of furnace temperature, pressure and air/fuel ratio?	☐	☐	☐	☐	☐
Heat recovery from furnaces & process plant?	☐	☐	☐	☐	☐

Electrical equipment

Power factor correction?	☐	☐	☐	☐	☐
Ensuring replacement motors are not oversized?	☐	☐	☐	☐	☐
Specification of high efficiency motors on motor replacement?	☐	☐	☐	☐	☐
Use of variable speed drives (VSD)?	☐	☐	☐	☐	☐
Automatic switch off of pumps, fans, conveyors & other equipment when not required?	☐	☐	☐	☐	☐
Purchase of energy efficient computers, photocopiers & other office equipment?	☐	☐	☐	☐	☐

11. Barriers to energy efficiency improvement

Studies by technology researchers commonly identify energy efficiency opportunities that appear to be highly cost effective. The following have been suggested as reasons why such investments are not made. In your view, how important is each suggested reason.

		Often important	Sometimes important	Rarely important
1	Technology inappropriate at this site	☐	☐	☐
2	Cost of production disruptions/ hassle/inconvenience	☐	☐	☐
3	Cost of identifying opportunities, analysing cost effectiveness and tendering	☐	☐	☐
4	Cost of staff replacement, retirement, retraining	☐	☐	☐
5	Possible poor performance of equipment	☐	☐	☐
6	Lack of capital	☐	☐	☐
7	Strict adherence to capital budgets	☐	☐	☐
8	Other priorities for capital investment	☐	☐	☐
9	Technical risk	☐	☐	☐
10	Business/market uncertainty	☐	☐	☐
11	Lack of information/poor quality information on energy efficiency opportunities	☐	☐	☐
12	Difficulty/cost of obtaining information on the energy consumption of purchased equipment	☐	☐	☐

13	Lack of time/other priorities	☐	☐	☐
14	Lack of technical skills	☐	☐	☐
15	Lack of staff awareness	☐	☐	☐
16	Department/individuals not accountable for energy costs	☐	☐	☐
17	Energy objectives not integrated into operating, maintenance or purchasing procedures	☐	☐	☐
18	Low priority given to energy management	☐	☐	☐
19	Energy manager lacks influence	☐	☐	☐
20	Conflicts of interest within the company	☐	☐	☐

Do you have any further comments on barriers to energy efficiency improvement?

....

....

....

Thank you very much for completing this questionnaire!

Please return this questionnaire in the enclosed pre-paid envelope

Annex B: Interview Protocol for Energy Manager

Background

- ☐ What are your principal ***responsibilities*** within the company?
- ☐ What proportion of your ***time*** is devoted to energy compared to other issues?
- ☐ How ***long*** have you been in this post?
- ☐ What is your ***professional background***?

Energy policy

- ☐ Does the company have an ***Energy/Environmental Policy***? What form does it take (e.g. use of targets; action plan; designated responsibility and accountability, etc.)?
- ☐ How ***long*** have these policies been established?
- ☐ What ***impact*** has this policy had on energy decision-making and the company's energy performance?
- ☐ What have the energy/environmental policies ***achieved***?
- ☐ Is the company certified to an ***environmental management scheme***? ISO 14000 or European EMAS? Does it intend to certify?
- ☐ Are provisions for achieving energy/environmental policy objectives included in other ***policy documents***, such as purchasing or maintenance?
- ☐ Is your company a ***subsidiary*** of another company? National? International? If yes, what impact does it have on energy management and decision-making? Do the same rules apply to all subsidiaries?

Energy management

- ☐ Could you please draw a simplified version of the corporate ***organisation diagram***, showing key individuals and departments/divisions with responsibility for energy decision-making.
- ☐ Does the company have a ***committee*** wholly or partly responsible for energy matters? If so, what is the ***membership*** of this committee? How does it function?
- ☐ Who do you ***interact*** with on a regular basis?

- ☐ Is energy management an in-house activity or are ***energy service companies***, consultants or energy suppliers used? What is the rationale for this choice?

Power

- ☐ How would you assess the ***status*** of energy management within the company (high/medium/low)?
- ☐ How does this status ***affect*** energy management activities?
- ☐ How much does status depend on ***formal authority*** and how much does it depend on other factors such as the support of key individuals (a lot / a bit / not at all)?
- ☐ Are there areas where you would like to have ***influence*** but do not (e.g. equipment purchasing, building specification)?
- ☐ Are there ***conflicts of interest*** within the company that inhibit energy management activities?

Investment

- ☐ Could you describe the ***decision-making process*** for investments in energy efficiency? Who are the key individuals/committees involved? What is the impact of the parent company (if applicable)?
- ☐ What are the biggest ***obstacles*** in this process? How could it be improved?
- ☐ What ***criteria*** are used for appraising energy efficiency investments? How do these differ from other categories of investment? What are the ***reasons*** for this choice of criteria?
- ☐ Are the investment criteria ***feasible*** or restrictive?
- ☐ Is there a ***portfolio*** of properly justified and costed projects that could be implemented?
- ☐ How are investment projects ***prioritised***? What factors are taken into account?
- ☐ Who has the final say?

Energy information systems (relate to answers in PIQ)

- ☐ What ***information*** is available to you on organisational energy consumption?
- ☐ To what extent is energy ***su- metering*** used in the company?
- ☐ Is information available on ***trends*** in energy consumption?
- ☐ Is information available on the performance of previous ***efficiency investments*** and the savings achieved?
- ☐ Is the energy performance of the company compared against sector or generic ***benchmarks***? Where are these used?
- ☐ Is a ***monitoring and targeting*** scheme in operation? How is it used?

- ☐ What information is disseminated to top management and other individuals? How effective are these ***reporting requirements***?
- ☐ Are you happy with the ***quality*** of current information systems? How could they be improved? What are the obstacles to such improvement?

Information on efficiency opportunities

- ☐ Do you consider that you have adequate information on energy efficiency ***opportunities***?
- ☐ Have any energy ***audits*** been conducted? Are any planned?
- ☐ Do you consider that you have adequate information on the ***needs*** of building and equipment users?
- ☐ Is the primary problem a ***lack of information***, or constraints on using existing information (e.g. lack of technical ability; lack of time)?
- ☐ Which ***information sources*** on energy efficiency opportunities are used and why? **(see PIQ)**
- ☐ Which information sources do you place the greatest ***trust*** in and why?
- ☐ Are you aware of any government sponsored ***information programmes***? How ***useful*** are they?
- ☐ How could the quality of information be ***improved***?

Accountability

- ☐ What are the arrangements for ***charging*** energy costs? Are subdivisions (e.g. brewing filling, packaging) charged individually?
- ☐ To what extent can individual subdivisions ***influence*** their own energy costs?
- ☐ Are energy budgets and the responsibility for energy management ***devolved*** to individual divisions?
- ☐ How are the ***benefits*** from efficiency investments distributed?
- ☐ Are ***targets*** for energy budgets identified?
- ☐ Are there any ***incentives*** created for subdivisions and staff? How effective are they?

Performance

- ☐ Do you know the energy ***savings*** achieved by your company over the past five years or so?
- ☐ Have energy savings measures been taken up for ***other reasons*** than to save energy (e.g. environmental reasons, such as emissions, odour).
- ☐ What types of energy efficiency ***measures*** have been implemented?
- ☐ How would you rate your company's ***performance*** on energy efficiency?

Capital

- ☐ Is there a separate ***budget*** for energy efficiency investment?
  - ☐ If yes: How large is it? (percentage of utility spend) How is it set? How has this changed over time?
  - ☐ If no: How is energy efficiency investment funded?
- ☐ How has this budget been ***spent*** over the last financial year?
- ☐ Is this budget allowed to be ***carried forward*** if unused at year end?
- ☐ How does capital availability affect the ***type of projects*** that can go ahead?
- ☐ Do you ***borrow*** for the purposes of investment in energy efficiency?

Overhead costs of energy management

- ☐ Which people are engaged in energy management activities? What are the estimated annual ***person-hours*** devoted to energy management?
- ☐ Do you know the estimated annual ***costs*** devoted to energy management activities?

Hidden costs

- ☐ Are the following considered when evaluating efficiency investments? How do you take account of them?
  - ☐ overhead costs of energy management
  - ☐ disruptions/hassle/inconvenience
  - ☐ cost of identifying opportunities, analysing cost effectiveness, tendering and seeking approval for expenditure
  - ☐ staff replacement, retirement, retraining
  - ☐ potential loss of benefit (e.g. unreliability, extra maintenance, **beer quality**)
- ☐ Have there been cases when additional costs were the reason for ***rejecting*** a project?
- ☐ Which ***technologies*** are particularly associated with such hidden costs?

Risk

- ☐ What, in your opinion, are the ***main risks*** facing your company (e.g. business economic trends; political decisions, takeovers)?
- ☐ What ***impact*** have these had on energy efficiency investment (e.g. through strict investment criteria)?
- ☐ Have there been cases where ***technical risk*** has inhibited the adoption of energy efficient technologies?

Equipment purchasing

- ☐ What influence do you have over ***equipment purchasing*** decisions? How does this vary between different types of equipment?
- ☐ In situations where you have an influence, what ***level of information*** is typically available on the energy performance of equipment? How difficult is it to obtain additional information on energy performance?
- ☐ In situations where you do not have an influence, do you think energy efficiency is considered? If not, why not?
- ☐ How important are written ***rules & procedures*** in purchasing decisions? Does energy efficiency feature in these rules?

Culture

- ☐ How important is ***environmental performance*** compared to cost saving in organisational decision-making on energy efficiency?
- ☐ Do you perceive any ***internal pressures*** to improve environmental performance (e.g. colleagues, students, etc.)? How important are these pressures?
- ☐ Do you perceive any ***external pressures*** to improve environmental performance (e.g. government, media, NGOs, local community, industry sector, etc.)? How important are these pressures?
- ☐ Is senior ***management aware*** of the potential for cost-effective investment in energy efficiency?
- ☐ Is senior management ***seriously committed*** to improving the environmental performance of the company? How widely is this commitment shared throughout the company?
- ☐ How important have ***changes*** in management and other key posts been in changing organisational performance on energy efficiency?

Awareness

- ☐ Have there been any actions taken regarding energy ***awareness campaigns***, energy training and incentive schemes? If yes, how effective have these been?

Perceptions of barriers

- ☐ Many studies have suggested that there are a large number of energy efficiency opportunities that are ***highly cost-effective*** at current prices. Do you think that this is the case with your company? If not, why not? If so, why are these opportunities not taken up?
- ☐ What do you see as the ***biggest obstacle(s)*** to improving energy efficiency in your company?
- ☐ Which energy savings opportunities are currently not taken up but could be profitably taken up in the future? (***wish list***)

Government Policy & ESCOs

- ☐ Do you consider ***contract energy management*** to be an attractive option for your company? If not, why not? If yes, then for which functions are they most appropriate?
- ☐ What are your views on the effectiveness of ***government policy*** on energy efficiency, as this impacts on your company?
- ☐ Do you benefit from ***governmental programmes*** (information, subsidies, tax breaks, etc.)?
- ☐ How should the ***government act*** to improve energy efficiency in the higher education sector? What do you consider to be the best ***instruments*** for improving energy efficiency?

Personal

- ☐ How ***long*** do you expect to remain in your current position?
- ☐ How is your ***performance*** in this position rewarded?
- ☐ Do you personally ***benefit*** from energy savings?
- ☐ What ***influence*** does this have on your choice of investment projects?
- ☐ Do you think that ***making money*** gets more recognition in this company than saving an equal amount of money?

Final

- ☐ Are there any ***important points*** about this topic which we have not discussed?

Index